# Recent Recruiting Trends and Their Implications for Models of Enlistment Supply

Michael P. Murray

Laurie L. McDonald

*Prepared for the*
*Office of the Secretary of Defense*
*United States Army*

*National Defense Research Institute • Arroyo Center*

RAND

Based on RAND's past body of recruiting research and on indications of increased difficulty in meeting recruiting goals, in spring 1994 the Army Chief of Staff and the Deputy Secretary of Defense asked RAND to examine recent trends in the recruiting market and their implications for meeting accession requirements. The request for assistance consisted of two parts: (1) a quick initial examination of the trends and (2) a longer-term research agenda to study the recruiting outlook in depth. The results of the preliminary examination were briefed in May 1994 and are documented in RAND report MR-549-A/OSD, *Recent Recruiting Trends and Their Implications: Preliminary Analysis and Recommendations* (Asch and Orvis, 1994). The current report presents some results from the longer-term analysis. In it, we update our econometric models of enlisted supply, and we use the results to determine if there have been important changes in the effectiveness of various recruiting inputs during the post–Cold War period. We also use the updated model to make predictions about the adequacy of supply in fiscal year 1997. The findings should be of interest to planners and policymakers concerned with recruiting.

Other reports produced as part of the larger research project include MR-677-A/OSD, *Military Recruiting Outlook: Recent Trends in Enlistment Propensity and Conversion of Potential Enlisted Supply* (Orvis, Sastry, and McDonald, 1996); MR-818-OSD/A, *Estimating AFQT Scores for National Educational Longitudinal Study (NELS) Respondents* (Kilburn, Hanser, and Klerman, 1998); and MR-845-OSD/A, *Encouraging Recruiter Achievement: A Recent History of Recruiter Incentive Programs* (Oken and Asch, 1997).

The research is being conducted within the Manpower and Training Program, part of RAND's Arroyo Center, and within the Forces and Resources Policy Center, part of RAND's National Defense Research Institute. The Arroyo Center and the National Defense Research Institute are both federally funded research and development centers, the first sponsored by the United States Army and the second by the Office of the Secretary of Defense, the Joint Staff, and the defense agencies.

# CONTENTS

# FIGURES

# TABLES

Concerns have arisen over the ability of the military services to meet their recruiting requirements over the next few years. The number of recruiters and the advertising budget have been cut over the course of the drawdown as recruit cohorts have shrunk by over a third. As compared with FY95, a 20 percent increase in accessions—greater than 40 percent for the Army—was needed by FY97 if the services were to meet the force strengths planned for that year.

Beginning in the early part of this decade, anecdotal evidence suggested that recruiters had been having trouble achieving their goals. A preliminary RAND investigation in 1994 indicated that there should not have been a supply shortage in the early-to-mid 1990s. Here, we revisit that question with a more thorough analysis, restricting ourselves to supply-projection models based on econometric analysis of specific supply and demand determinants. (Other parts of the larger research project examine other aspects of recruiting trends.) As the parameters in earlier models were estimated with data from the 1980s, we reestimate the models with 1990s data to determine whether those parameters—and thus the underlying supply process—have changed. We also determine whether any changes in factors influencing supply occurred. Finally, we use the updated model using 1990s data to predict the adequacy of supply in FY97.

To compare models estimated with data from the 1980s and 1990s, we needed comparably structured models and comparable data. The easiest way to set up the comparison was to create models—one for each service—and estimate them for both 1980s and 1990s data.

Our econometric models relate the number of high-quality non-prior-service contracts in a given month in a multicounty area to a set of variables representing youths' opportunities and the military's recruiting efforts. The former include civilian unemployment indicators and civilian pay relative to military pay. The latter include relations between goals and achievement, along with the number of recruiters in the area and high-quality youths available (by race and ethnicity). The effect of advertising on high-quality recruits is estimated for each service, and the effects of enlistment bonuses and the Army College Fund are estimated for the Army.

Elasticities for the new models based on 1980s data were generally not far off from those for the old models used in the preliminary investigation; the estimated effect of advertising was the sole exception. And the results were similar; the new service-specific models estimated with 1980s data (except that for the Marine Corps) overpredicted the number of high-quality contracts in the early 1990s, as the original models devised in the 1980s had done. Thus, we confirm the results of our preliminary investigation.

We compared the coefficients of the models estimated with the 1980s data to those estimated with the 1990s data. Most of the comparisons yielded statistically significant but not practically important differences. The exception was the result for recruiters. We find that the impact of recruiters was smaller in the 1990s period relative to the 1980s period for the Army and the Air Force. In other words, we find that the number of additional high-quality contracts produced by adding recruiters fell between the 1980s and 1990s for these two services. The reason for this result is not entirely clear. It may be due in part to a decline in youths' interest in military service or to important changes in recruiting management and resource allocation as resources were cut during the drawdown. These explanations are explored in other reports produced under this research project.

Finally, we find that our updated models predict that the Army would have difficulties meeting its FY97 recruiting mission. To address the recruiting difficulties it actually did face in FY97, the Army increased the Army College Fund benefit amounts, the enlistment bonus benefit amounts, and its advertising budget during FY97. The Army also reduced its accession mission. As a result, the Army was able to meet its mission in FY97.

# ACKNOWLEDGMENTS

We would like to express our gratitude to the Office of the Army Deputy Chief of Staff for Personnel, and OSD's Accession Policy Directorate, our sponsors. We are also grateful to the Defense Manpower Data Center, and in particular to Dr. Robert Tinney for providing data and much helpful counsel and to Mr. Les Willis and Mr. Neil Weil for providing the contract data and insights into their use. Further thanks are due the U.S. Army Recruiting Command for its cooperation in providing recruiting information. We also thank our RAND colleagues Beth Asch, James Dertouzos, Glenn Gotz, Rebecca Kilburn, and Jacob Klerman for fruitful discussions about enlistment supply modeling; Carole Oken for data support; Sally Carson for providing census data; and James Chiesa for carefully reading and commenting on this document. We also thank James Hosek for his thoughtful technical review. Finally, we are indebted to Madhur Duggar and Seth Murray for outstanding research assistance, and to Nikki Shacklett for her careful editing.

# Chapter One

# INTRODUCTION

A steady flow of bright, well-educated young men and women is the lifeblood of the American military. If able new recruits do not step forward to serve their country, the armed forces cannot sustain their readiness. In 1994, OSD and the Army approached RAND about several concerns the military had about the flow of new recruits. These concerns had to do with lowered resource availability for military recruiting, a reported reduction in youths' interest in joining the military, and scheduled increases in military accession requirements in upcoming years.

Between FY89 and FY94, the number of Army recruiters was reduced 25 percent and recruiting advertising expenditures were cut by 50 percent. These resource cuts had been made in large part because in the post–Cold War military drawdown, accession requirements had been sharply curtailed. In FY89, the last predrawdown year, annual non-prior-service military accessions were about 275,000. By FY94, the number of accessions had shrunk to 175,000. Nonetheless, the services and OSD were concerned that the resource cuts might have been too large. Reports that recruiters in the mid-1990s were experiencing increased difficulty in meeting their monthly accession goals, coupled with widespread reports that youths were becoming less inclined to join the military, fueled these concerns. Another worry was the increase in military accession goals scheduled to take place beginning in FY95. By FY97, military accessions were to rise by nearly 20 percent above their FY94 levels; the increase for the Army was to be nearly twice that large.

In 1994, OSD and the Army asked for RAND's help in assessing recent recruiting trends and their implications for meeting accession requirements. This report is one element of that assessment. Here we ask if past econometric models based on data from the early-to-mid 1980s predict the recruiting difficulties reported in the mid-1990s, and, if not, whether new models are needed. We also ask what a model appropriate to the 1990s tells us about the adequacy of high-quality recruit supply in FY97. For each service we examine the effects of unemployment, pay, recruiters, advertising, and the qualified youth population. For the Army, we further consider the effects of enlistment bonuses and the Army College Fund.

RAND first reported its project findings on recruiting trends in Asch and Orvis (1994). Asch and Orvis found that the supply of potential enlistees exceeded its predrawdown level relative to the accession requirement. They argued that reported problems in meeting monthly recruiting goals could be due to changed recruiting practices that might have resulted in difficulties in converting potential supply into enlistment contracts. Because both their econometric and attitudinal models yielded consistent estimates of enlistment supply, it was unlikely that their estimates of the enlisted supply markedly overpredicted the actual levels. Nonetheless, the project team has made it part of its overall assessment strategy to reexamine the efficacy of the models used in the preliminary assessment. For example, Orvis et al. (1996) reports the assessment of the attitudinal models. The current report provides an analysis of the econometric models of high-quality enlistment supply.

## HIGH-QUALITY ENLISTMENT SUPPLY AND DEMAND

Potential recruits to the military face a choice among further education, the civilian work force, working at home, and enlisting in the military. Potential recruits balance the advantages and disadvantages of each alternative to choose the most attractive life choice for themselves. Each alternative offers its own package of pecuniary and nonpecuniary rewards and costs, about which individuals have imperfect information. These rewards and costs constitute the "supply determinants" of high-quality enlistment. Chief among them are civilian pay opportunities, military pay and benefits (e.g., the Army

College Fund), unemployment risks in the civilian sector, and the costs of further education. Because the perception of civilian opportunities and costs differs among individuals, different demographic groups may have different propensities to enter the military; consequently, demographic variables, such as race and ethnicity, may also enter the supply relationship. Early econometric models of enlistment supply (e.g., Ash, Udis, and McNown (1983) and Brown (1985)) focused exclusively on the role of such supply determinants.

Accession requirements are filled by a mix of high-quality recruits and others. Consequently, the number of high-quality recruits who join the service in any period is the number who wish to join, not the number the military would like to see joining; that is, we always observe the quantity of high-quality recruits supplied, not the quantity demanded. However, this is not to say that demand-side considerations do not influence the supply of high-quality recruits.

Because potential recruits have imperfect information about the military, the behavior of recruiters and job counselors as well as military advertising can influence the decisions of potential enlistees by improving their information about the opportunities in the military. In practice, such information's role has proved to increase recruiting. These effects of recruiters, job counselors, and advertising have been documented in past RAND research (e.g., Fernandez (1982), Dertouzos (1985), Polich, Dertouzos, and Press (1986), Dertouzos and Polich (1989), Asch (1990), and Asch and Karoly (1993)). Moreover, recruiting practices, such as restrictions on the numbers of low-quality recruits and other policy mechanisms, also influence the effort given to attracting high-quality recruits. Consequently, the supply of high-quality enlistees will be influenced not only by the traditional "supply determinants" noted above but also by "demand factors" operating through the efforts of recruiters and counselors. Chief among these demand factors are the number of recruiters, the goals they are given, the recruiters' expectations of success in meeting their goals, incentives given to recruiters and counselors, and various forms of advertising expenditures. The econometric models we consider here are those that incorporate both supply determinants and demand factors that influence the supply of high-quality enlistments.

## PLAN OF THE REPORT

Our analysis proceeds in two stages. First, we estimate econometric models of enlistment supply for each of the four services, allowing for the possibility that the underlying structure of supply may have changed with the end of the Cold War or the drawdown of U.S. military forces. We formally test whether such a change occurred, using conventional measures of statistical significance. Second, we use models appropriate to the 1990s to ask whether the supply of high-quality recruits would be adequate in fiscal year 1997.

In our initial examination of recruiting trends, we relied upon elasticity estimates drawn from several econometric studies that were conducted during the 1980s. With those estimates we were able to show that past econometric models did not predict the reported decline in high-quality enlistment supply for the Army.

However, the old estimates alone are insufficient to determine whether there has been a change in the underlying econometric model that should be used for forecasting enlistment supply. We need estimates of the elasticities based upon recent data for comparison. Moreover, the two sets of estimates compared must be based upon comparable data sets, using comparable methodologies. Thus, we must not only assemble data for estimating the model in the new time period, we must assemble a data set suitable for estimating the econometric model in both time periods. Comparison of the elasticity estimates across the two time periods allows us to decide whether or not the underlying econometric model has shifted.

Statisticians have established formal criteria for determining when the estimated coefficients in a model differ enough to warrant calling them "statistically different." These criteria are met when it is highly unlikely that the coefficients would by chance differ by as much as they appear to. "Statistical significance" differs from "economic significance." The difference between estimated elasticities of 0.90 and 0.91 may prove "statistically significant," but the practical consequences of such a difference may be so small as to make the difference "economically insignificant." We examine both the statistical and economic significance of differences we see between models estimated for the earlier and later time periods.

This report is organized as follows.  Chapter Two discusses the variables that appear in the econometric model; formal theoretical justification for the model is relegated to Appendix A.  Chapter Two also describes the data used in this report.  Chapter Three briefly describes the econometrics of estimating high-quality enlistment supply; Appendix B contains a more elaborate discussion of the econometrics.  Chapter Four reports the empirical findings for each of the four services (Army, Navy, Air Force, and Marines).  Chapter Five presents the conclusions.

# VARIABLES INFLUENCING ENLISTMENT SUPPLY

## OVERVIEW

Early econometric models of enlistment supply (for example, Ash, Udis, and McNown (1983), Goldberg (1982), Dale and Gilroy (1985), Brown (1985), and a summary by Nelson (1986)) focused exclusively on the number of potential recruits and on variables that reflected the relative attractiveness of military and civilian jobs. But how many high-quality youths sign up to join the armed forces also depends on their attitudes toward the military. An example of a study of this dimension is the work by Dertouzos and Polich (1989), which emphasizes the role of military advertising and its effects on youths' willingness to serve. Still other analyses recognized that recruiters' effort levels would influence the number of high-quality recruits, and that such effort would in part be determined by the resources devoted to recruitment (for example, Dertouzos (1985), Daula and Smith (1985), Berner and Daula (1993), and Polich, Dertouzos, and Press (1986)) and the incentives given to recruiters (see Asch (1990) and Asch and Karoly (1993)). This report builds on these earlier efforts. The models estimated here presume that the supply of high-quality recruits is affected by advertising and the intensity of recruitment, as well as by traditional supply variables such as pay, bonuses, and benefits.

Analysts have differed in the kinds of data they use to study enlistment behavior. Hosek and Peterson (1985, 1990) have used individual-level data. Many other analysts have used aggregate data to study enlistment supply, with aggregates sometimes defined across the entire country for a given time period, sometimes for individual states for a given time period. We aggregate enlistments for each of hundreds of small geographic areas in the country, each month over a period of several years, and each military service. Most analysts have used time-series data to study enlistment behavior; some have used cross-sectional data. Still others have used a combination of cross-sectional and time-series data, which, as just noted, is the approach we take in this report.

In this chapter we describe the variables that will appear in our econometric models, providing rationales for their inclusion in the models and information about data sources. Before turning to that discussion, however, we provide a brief overview of the time-series/cross-section structure of our data.

## TIME-SERIES/CROSS-SECTION STRUCTURE

The data used here are time-series of cross-sections for each of the four services; the unit of observation is a given geographic area in a given month. The data are monthly, spanning fiscal year 1983 (FY83) through FY93, the last year for which net contract data were available when we created our files. The cross-sectional unit of observation is the public-use microdata area (PUMA), an aggregation of counties defined by the Census Bureau. PUMAs cluster counties with small populations into entities large enough to be better suited to statistical analysis. We group the data for every month by 1990 PUMA definitions.[1] We have complete data for 911 PUMAs; we dropped 35 PUMAs from our analysis for lack of complete information. In constructing our data files, we first created a file at the county level, which we then aggregated to the PUMA level before conducting the analysis. We restrict our analysis to the contiguous United States.

By using county-level (or, more precisely, PUMA-level) data, we increase the variability in the observed values of several variables relative to that obtainable with the more aggregated geographical areas used in previous studies. That is particularly true of unemployment rates and population counts of qualified youths. Such increased

---

[1] Several very populous counties are actually broken into several PUMAs. We have no data below the county level, so we could not study these PUMAs individually. For them we include their county counterparts in our analysis file. Nevertheless, we describe all the data as observations on PUMAs.

variability promises more precise estimates of the parameters of our econometric models than would otherwise be possible. Previous analyses have not had access to the array of county-level data that have become available in recent years. Consequently, some studies have focused on state-level data, and others have used military recruiting unit areas as their cross-sectional unit of analysis. Appendix C compares our empirical results with those we obtain from analyzing the data aggregated into military recruiting units.

## VARIABLES LINKED TO YOUTHS' OPPORTUNITIES

### Civilian Opportunities

High school graduates who enlist in the military forgo either civilian job opportunities or immediate further schooling. The more attractive those alternatives, the less likely it is that graduates will join the military. Two variables are commonly used to capture the attractiveness of civilian opportunities: unemployment and the average level of civilian pay opportunities. High school graduates who enter the civilian labor force face the risk of becoming unemployed. The higher that risk, the less attractive entering the civilian labor force will be. On the other side of that ledger, the higher the pay for civilian work, the larger the number of graduates who will enter the civilian work force. Past studies have not included measures of the attractiveness of further education for high school graduates. We do not overcome this inadequacy here.

We have two measures of unemployment. The first is monthly counts of unemployment by county provided to us by the Defense Manpower Data Center (DMDC); these counts were prepared for DMDC by the Bureau of Labor Statistics (BLS) from a variety of sources. We use this first measure in our basic econometric model. The second is state-level average annual unemployment rates by race, drawn from Department of Labor publications. We combine the state-level unemployment rates by race with the estimated numbers of high-quality high school graduates (discussed below) to form estimates of the number of unemployed high-quality high school graduates in each PUMA. The two series follow similar patterns; we rely on the BLS measure as the better proxy because it should better capture PUMA-to-PUMA variation in employment opportunities.

We also have two measures of civilian pay opportunities. The first is the Defense Employment Cost Index (DECI), a nationwide annual index of civilian pay for people similar to those employed by the military; the development of the DECI is described in Hosek, Peterson, and Heilbrun (1994). The second is drawn from the March Current Population Survey. This second measure is the smoothed annual average earnings for a random sample of civilians with a high school diploma (but no college) who have been in the labor force at least 40 weeks in the preceding year. These earnings data were compiled by the Center for Naval Analyses, and given to us by DMDC. As the DECI varies only at the national level, we choose to use the earnings data, which vary at the state level, in our econometric analyses, combining them with the military pay index described in the next section to form a ratio of civilian to military pay opportunities.

## Military Opportunities

The attractiveness of enlistment depends upon military pay and fringe benefits (including any bonus or college fund programs for which an individual is eligible). To measure military pay for each of the four services, we use the Basic Pay Index (BPI). The BPI does not include the basic allowances for pay, quarters, and subsistence, but according to Hosek, Peterson, and Heilbrun (1994), percentage increases in the BPI and these allowances have been equal for every year but one since 1982. The BPI thus serves well as an index for total military pay.

The BPI also changes only annually and only at the national level, which limits its value as a predictive variable for our purposes. However, potential enlistees only care about the relative generosity of military and civilian pay, not their individual levels. Consequently, the variable that belongs in an econometric analysis is the ratio of civilian to military pay, rather than the two variables separately.

Enlistment bonuses and the Army College Fund (ACF) offer further financial incentives to potential enlistees. Not all enlistees are eligible for these programs, and who is eligible has varied widely over time. Because the bonus and college fund programs are uniform across the country, we observe variation in these programs only over time. Consequently, the amount of variation we observe in these programs is quite limited, occurring only with the quarterly changes in eligible specialties and benefit amounts. (Indeed, the benefit offered by the Army College Fund has changed very seldom over the years.) This limited amount of information presses us to seek a parsimonious description of the programs for use in our econometric model. After some experiment we chose to use similar measures for both programs, namely the mean of the bonus amount offered across all occupational specialties discounted into current dollars at the time of enlistment. This measure is sensitive to both the number of specialties being offered opportunities and the magnitude of the benefit offered.

The ACF poses a special problem in that the benefit is not realized by participants immediately, but only when they attend school. We rely on the present discounted values of ACF benefits calculated in System Research and Applications Corporation's Army College Fund Cost-effectiveness Study to provide a measure of the value of ACF benefits for use in this study.

Recruitment advertising alerts potential enlistees to the merits of military service. We attempt to include in this study an analysis of the effects of advertising expenditures on high-quality recruit supply for each of the services. Unfortunately, the only measure of advertising available to us is the total annual expenditure on recruit program advertising. This measure proves unsuitable for successful econometric analysis of advertising's effects on recruit supply. The impact of advertising is not uniform across the United States. Some geographical areas are reached far more intensely by advertising expenditures than others are. We had no data to capture this dimension of advertising's effects. With only annual national advertising expenditures, we have too little data to see advertising's effects; moreover, what effects we do see are probably contaminated by advertising's correlation with other, unmeasured national policies. The omission of advertising is particularly unfortunate because the level of recruitment advertising was much higher in our earlier sample period than in the latter and could potentially account for the increased difficulty in recruiting in the later period.[2]

---

[2]Another RAND research project is investigating the effects of advertising.

The attractiveness of military service depends on the current missions of the military as well as on compensation. For example, in the 1960s and 1970s, the military was fighting in Vietnam. This influenced enlistment decisions. More recently, the end of the Cold War has altered the role of the military in international affairs. No longer is the military serving to contain "the Evil Empire." Instead, it has become an instrument of government policy in a number of regional combat and peacekeeping missions. Military service may seem a very different thing today than it did during the Cold War, so the structure of enlistment decisions may have altered radically with that era's end. Indeed, such a fundamental shift in youths' attitudes toward the military is one possibility that makes it necessary to ask if econometric models using 1980s data can accurately forecast 1990s enlistment behavior.

Competing with the hypothesis that attitudes have fundamentally changed are two alternatives: (i) a claim that there has been no fundamental change in enlistment supply and (ii) a claim that there has been a change in supply that is linked not to the end of the Cold War, but to the drawdown of the military that its end engendered. To allow for all of these possibilities in our analysis, we choose to focus our attention on two periods: the Cold War years FY83–FY87 and the drawdown years FY90–FY93. We drop both FY88 and FY89 from our analysis, because which period they belong to—or whether they should be treated as a distinct third period—depends critically on the very issues we are asking about. One further benefit of our using PUMA-level data is that even when we drop these 20 percent of our observations, we continue to have an ample sample size for our investigations.

## VARIABLES LINKED TO RECRUITING EFFORT

Conventional measures of enlistment supply, those reflecting youths' opportunities, determine the "potential" pool of military recruits. But those potential recruits must be transformed into enlistees through the efforts of recruiters. Dertouzos (1985) was the first to formalize the relationship between recruiter effort and enlistment supply. We follow the broad outlines of his approach.

Dertouzos argued that recruiters' utility depends on how many low-quality and high-quality recruits they sign, the contract goals they are given, and how hard they work. He notes (p. 4) that "recruiters do not passively process enlistments; rather, they have considerable discretion over the allocation of resources, the most important of which is their own time." Time can be spent in many different ways, for example, at high school "career day" programs, at Eagle Scout gatherings, at shopping malls, or in the recruiting office waiting for "walk-ins." Some of these activities would better enhance the signing of high-quality recruits, others the signing of low-quality recruits. Still another alternative would be for the recruiter to spend more time at home with his or her family.

How the recruiter divides time among seeking high-quality recruits, seeking low-quality recruits, and pursuing personal ends depends on the incentives the recruiter associates with each alternative. Commanders signal the absolute importance of contracts by setting goals for recruiters; commanders have sometimes further signaled the relative importance of high- and low-quality contracts by assigning separate goals for each. Since promotion rates are likely to be influenced by recruiters' success in meeting (and exceeding) goals, recruiters value signing recruits to contracts, and how much they value this is influenced by their contract goals.

Goals are not the only mechanism the services use to give recruiters an incentive to work harder. Asch (1990) and Asch and Karoly (1993) examine how incentive plans, such as the Navy's Freeman Plan, influence recruiter effort. We do not include such effects in our models here.

The recruiting success of one service may depend upon the efforts of other services, though the direction of such cross-service effects is not entirely clear. For example, an increase in Army recruiting effort may lure away potential Navy recruits or it may spur Navy enlistments by making the military option more apparent to potential recruits. We explored including such cross-service effects in our models but were not able to estimate them with sufficient precision to shed light on this question.

## Expected Recruiting Performance

Dertouzos simplified his general formulation by assuming that recruiters cared only about how many recruits they signed relative to their contract goals.  He assumed that when recruiters expect to perform well relative to their assigned goals, they work less hard than when they expect to face difficulty in meeting their goals.  We, too, let effort depend on recruiters' expectation of success.

It is because recruiters make their decisions about how hard to work before they know how many recruits they will sign that they need to form some expectation about how many recruits they will sign.  Unfortunately, recruiters' expectations are unobserved, so we must choose a proxy variable to mimic their expectations.  How are we to do this?  Early in a month, recruiters can only look to past experience to set their expectations; later in the month, they can see how many contracts they have already obtained.  Thus one could choose either past or present performance to proxy for recruiters' expectations (the two are too highly correlated to make using both an attractive prospect).  Dertouzos chose to use the actual contracts and goals in a period to proxy for expectations; we choose to use the previous period's actual contracts and goals.  In practice, the two measures move closely together, but we prefer our measure because it simplifies the econometric problems one faces in estimating a model of high-quality enlistment supply.  In our linear specification, we use the difference between a recruiting unit's contracts and goals (in the previous month) to measure expected relative success, so the variable that appears in the model is this difference.  (In his logarithmic specification of enlistment supply, Dertouzos used the difference in the logarithms of contracts and goals to measure this expected relative success, which is the same as using the logarithm of the ratio of contracts to goals.)

In the Army, recruiters are given both low- and high-quality goals. Like Dertouzos, we expect that recruiters who expect high performance in meeting their low-quality goals will feel free to devote more work effort to garnering high-quality enlistments.  In preliminary analyses we included a measure of performance in meeting low-quality goals in our model of high-quality Army enlistments.  The estimated signs on low-quality performance and on low-quality goals were always as expected, but the estimates were never statistically significant, so we have excluded them from the analysis reported here.

## Goals

Dertouzos assumed that the effect of goals on enlistments was fully captured in the ratio of a unit's contracts to its goals. We, on the other hand, allow the level of goals to have a separate influence on recruiters' effort, with the expectation that if two recruiters have the same expected relative success in meeting their goals, the recruiter with the higher absolute goal will expend more recruiting effort.

DMDC has provided us with quarterly goal data for each of the four services, at the recruiting-unit level. (In the Army the recruiting unit is a battalion; in the Navy it is a district; in the Air Force it is a squadron; and in the Marine Corps it is a station.) We then determined a crosswalk between census PUMAs and recruiting units to match goals to PUMAs. (In some cases, a PUMA was served by several recruiting units; then we averaged the goals of all the units, weighing each by the size of the population it served.) These data, coupled with the contract data described below, allow us to calculate for each PUMA, in each month, the difference between contracts and goals in the previous month.

The difference between a unit's contracts and its goal (lagged one month) is our performance variable, as noted in the previous subsection. Across our entire sample period, the Army, Air Force and Marine Corps assessed recruiters' performances relative to their individual goals, not unit goals; the Navy followed this same practice in the earlier period, but for much of the later period it assessed recruiters based upon their unit's overall performance. Since our contract and goal data are at the unit level, it is more accurate to think of our performance measure as being pertinent to the unit commander, an individual whose performance ratings are always based upon the unit's overall success, rather than pertinent to the individual recruiters, but we abstract from this distinction in our discussion below.

Although each service has long set contract objectives by quality, the Air Force, Navy, and Marine Corps have not set quality goals for recruiters across the entirety of our sample period; for example, in our data the Navy began setting quality goals in addition to overall goals late in our sample period. Consequently, we do not analyze high- and low-quality goals separately for the Navy and Air Force. Instead, for these two services we use a measure of performance based upon total contracts and total goals. DMDC reported that the goal data for the Marine Corps were incomplete and not suitable for analysis, so we do not include either performance or goal in the Marine Corps enlistment supply model. For the Army, we use high-quality contracts and goals to measure performance.

## Recruiters

Recruiting units with larger numbers of recruiters can obviously expend more overall effort seeking recruits. We therefore include in our models a count of the number of production recruiters working in a geographic area. The specific measure we use was provided by DMDC. Recruiters are assigned to recruiting units, not counties or PUMAs. To assign recruiters from recruiting units to counties, DMDC identified all the counties served by a particular recruiting unit (each county is served by only a single recruiting unit) and apportioned the recruiting unit's recruiters across those counties in proportion to their populations of 17–21 year olds. We simply aggregated these counts of recruiters in a county to the PUMA level.

## Numbers of Youths and Numbers of Contracts

The number of high-quality enlistees will depend in part on the size of the pool of high-quality high school graduates the military can tap. Some past studies have used census counts of population as proxies for the number of high-quality high school graduates, for example the number of males aged 17–21. We instead use county-level estimates of the number of high-quality high school graduates provided to us by DMDC and developed at the Naval Postgraduate School (Thomas and Gorman, 1991); these population counts are divided into three variables: the total estimated number of high-quality youth in the PUMA, and such counts for black and for Hispanic high-quality youth. (DMDC provided estimates, by race and ethnicity, of both the number of high school graduates and the number of high school graduates who would score in categories I–IIIA on the AFQT).

For each service, DMDC has provided us with monthly counts of net and gross non-prior-service enlistment contracts, by county (we have combined males and females to tally total non-prior-service contracts).  The DMDC counts have the virtue of being based upon an unvarying definition of "net" over the entire sample period, i.e., those contracts signed by individuals who are still in a service 12 months after signing up.  Our initial intent was to study net contracts, this being the more important figure for the services.  However, we examined gross contracts instead because we found data problems in the net contracts series (see Figures 1–4).  First, the gross and net contract figures for FY88 appeared far too low (DMDC has confirmed that transitions in data management and reporting difficulties in FY88 make those data thoroughly unreliable).  Second, the net contract data for FY92 appeared too low; DMDC is analyzing these data to determine the source of this problem.  Fortunately, while Figures 1–4 make clear the difficulties in the net contract data, they also make clear that in years other than FY88 and FY92, gross and net contracts move very closely together.  Consequently, a change in the structure of a model of net contracts would be reflected in a comparable model of gross contracts.  To allow us to use data from FY92, we chose to analyze gross contracts instead of net contracts.[3]

The difference between gross and net contracts arises because most enlistees do not enter the military on the day they sign the enlistment contract.  Instead, most individuals sign an enlistment contract in one month, say January, but do not actually enter the military until a later month, such as June.  Many enlistees actually enter the military in the summer after they graduate from high school, though they signed enlistment contracts at some earlier date.  When an individual signs the contract, he is in the Delayed Entry Pool (DEP).  He exits the DEP when he actually enters service.  Not every individual who signs a contract actually enters service.  Because new opportunities arise, minds change, or unforeseen events occur, some individuals attrite from the DEP—they sign a contract but do not enlist.  To account for DEP attrition, the recruiting commands focus on both gross contracts and net contracts.  Net contracts equal gross contracts minus those individuals who attrite.  For example, if 100 individuals sign a

---

[3]The data from FY88 are unusable in any case. We did analyze the net contract data with neither FY88 nor FY92 in the data set, and found little qualitative difference from the results reported here.

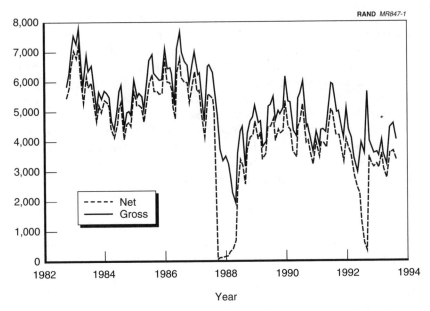

**Figure 1—Army Net and Gross High-Quality Contracts, by Month**

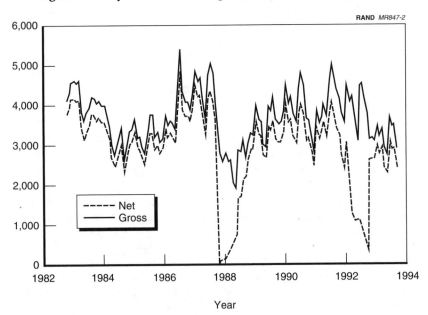

**Figure 2—Navy Net and Gross High-Quality Contracts, by Month**

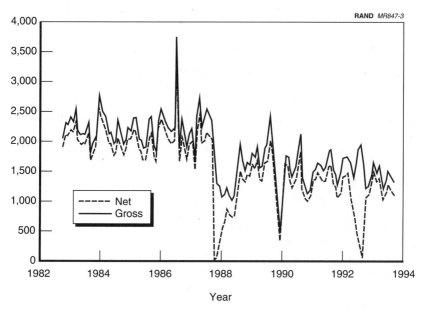

**Figure 3—Air Force Net and Gross High-Quality Contracts, by Month**

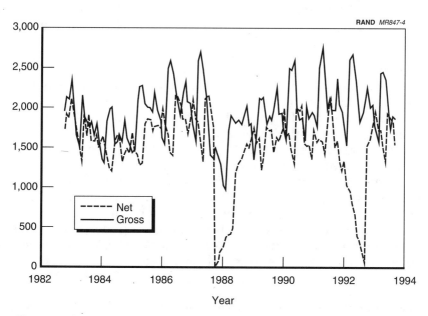

**Figure 4—Marine Corps Net and Gross High-Quality Contracts, by Month**

contract in November but 10 individuals later attrite from the DEP, net contracts equals 90.

## Geographic and Temporal Dummy Variables

The richness of career and schooling opportunities available to high-quality high school graduates varies widely from place to place, and these differences are only partially captured by earnings levels and unemployment rates. Attitudes toward military service also vary widely from place to place. To capture some of this geographic variation in opportunities and attitudes, we include in our model dummy variables indicating a PUMA's state.[4]

Similarly, the recruiting policy environment varies considerably over time in ways not captured by military pay, goal levels, bonuses, or college benefits. If we take no account of these unmeasured policy shifts, we risk obtaining biased estimates of the coefficients in our models. One way to take account of these policy shifts is to incorporate temporal dummy variables into our models, dummy variables that indicate which particular year or month we are observing. Because our PUMA-level data give us many observations in each time period, we are able to include dummy variables for each month into our analysis of many variables' effects on recruiting.

Unfortunately, several important policy variables (advertising, bonuses, and college fund benefits) are constant across the nation and within time periods. The effects of these variables cannot be estimated if there are dummy variables for each month (or even each year) in the model—these policy variables are perfectly collinear with the temporal dummies. We therefore conduct our analysis in two steps. First, we estimate all the coefficients we can in a model with the temporal dummies included; these coefficients are estimated with minimal risk of bias. Second, we estimate the effects of advertising, bonuses, and college benefits by replacing the temporal dummies in the model with variables measuring each; these latter estimates unavoidably risk bias, but they are the best available estimates of the effects of these three policy variables.

---

[4]Initially we allowed each PUMA its own dummy variable, but statistical testing showed this added no information beyond that captured by state dummies.

# Chapter Three

# THE ECONOMETRIC MODELS

The central insight underlying our econometric models of enlistment supply is that the observed supply of high-quality recruits depends both upon traditional supply factors that capture the determinants of potential enlistments, and upon the recruiting effort expended to attract potential recruits into the service. However, since we cannot directly observe recruiting effort, an econometric model of enlistment supply must capture its effect indirectly, through a series of variables that reflect the essential determinants of recruiter effort.

In this chapter we first present our formal econometric model and highlight what we can and cannot learn from it. We then briefly discuss our estimation strategy.

## A FORMAL MODEL

The following model captures the complexities of enlistment supply noted above. The model is liberally adapted from that used in Polich, Dertouzos, and Press (1986). Appendix A presents a more elaborate development of the model.

The supply of high-quality recruits is specified as

$$H = \gamma_1 D + \gamma_2 S + \gamma_3 X_H + \gamma_5 E_H + \varepsilon_H, \tag{1}$$

in which $H$ is the number of high-quality recruits, $D$ is a vector of time-period dummy variables, $S$ is a vector of dummy variables indicating state, $X_H$ is a vector of traditional determinants of high-quality supply (including civilian and military pay, unemployment, a measure of population size, military advertising, and, where applicable, measures of bonuses and college fund opportunities offered to recruits), and $E_H$ is a measure of recruiting effort devoted to garnering high-quality recruits. Since recruiter effort, $E_H$, is unobserved, some adaptation of this specification is needed before the model is estimable. The term $\varepsilon_H$ is for stochastic disturbance.

Following Polich, Dertouzos, and Press (1986), we assume recruiters make choices between expending effort to recruit high- and low-quality recruits and enjoying leisure time, in accordance with the incentives they face. Implicit in Dertouzos (1985) was the fact that recruiters make their decisions about work effort before they know how many contracts they will garner. Consequently, recruiters' choices of effort level depend upon their expectations of what their contract performance is likely to be relative to their assigned goals.

As mentioned in Chapter Two, Polich, Dertouzos, and Press (1986) chose to use current contract achievement as a proxy for the recruiters' expectations. (Doing this facilitated some of that study's mathematical analyses.) We choose to use lagged contract achievement because it simplifies the econometric analysis needed to estimate the model. (For further discussion of this simplification, see Appendixes A and B.)

Polich, Dertouzos, and Press (1986) assumed that recruiters' effort depends only on their expected performance relative to assigned goals. We think it more realistic to believe that recruiters' effort level will also be influenced by the specific goals they are given and on how difficult or easy it is to attract recruits in a given recruiting area. Finally, we think the level of recruiting effort will also depend on the number of recruiters working in an area. These considerations lead us to the following formal model for $E_H$:

$$E_H = \tau_1 + \tau_2 X_E + \tau_3 M_H + \tau_4 M_L + \tau_5(P_H) + \tau_6(P_L) + \tau_7 R + \varepsilon_1, \qquad (2)$$

in which $X_E$ contains numerous influences that affect recruiters' choices of effort level. The elements of $X_H$ are obvious candidates for inclusion in $X_E$, since they are indicative of how hard or easy it is to attract recruits; variables capturing incentive plans like the Navy's

Freeman Plan might also be included in $X_E$, but we do not do so here.[1] $M_H$ and $M_L$ are the recruiters' high- and low-quality recruiting goals; $P_H$ and $P_L$ are the recruiters' performance last month vis-à-vis high- and low-quality goals (the differences between contracts and goals last month); $R$ is the total number of recruiters, and $\varepsilon_1$ is a stochastic disturbance term. When recruiters got more recruits than their high-quality goals last month ($P_H > 0$), we expect them to slack off some from high-quality recruiting effort and to possibly focus more on low-quality recruiting, and contrariwise when they exceeded their low-quality quota ($P_L > 0$). Hence, we expect $\tau_5$ is negative and $\tau_6$ positive. When high-quality goals are high, we expect more effort to be expended on garnering high-quality recruits; when low-quality goals are high, we expect effort to be drained away from seeking high-quality recruits. Consequently, we expect $\tau_3$ is positive and $\tau_4$ is negative.

The definitions of $P_H$ and $P_L$ require some discussion. Our units of observation are at the PUMA level. But recruiters' success is not defined at the PUMA level. Rather, recruiters are judged by how they perform at the recruiting-unit level. Hence, $P_H$ and $P_L$ are defined in terms of the number of contracts signed last month in the recruiting unit that serves this particular PUMA.

Since effort is unobserved, neither (1) nor (2) is estimable. However, by substituting (2) into (1) in lieu of $E_H$, we obtain an estimable supply equation for high-quality recruits:

$$
\begin{aligned}
H = {} & \beta_1 + \gamma_1 D + \gamma_2 S + \gamma_3 X_H + \beta_2 X_E + \beta_3 M_H + \beta_4 M_L \\
& + \beta_5 (P_H) + \beta_6 (P_L) + \beta_7 R + u_H,
\end{aligned}
\tag{1a}
$$

where $\beta_i = \gamma_5 * \tau_i$ and $u_H = \varepsilon_H + \gamma_5 \varepsilon_1$. (1a) is the basic econometric model of high-quality recruit supply that we use in this report. We rely on a linear specification rather than the logarithmic specification that has commonly been used in the past. We do this because for many PUMAs, no contracts are signed in some months, making a logarithmic specification impossible, since the logarithm of zero is undefined. (Appendix C reports the results of a logarithmic analysis

---

[1] In our estimations, we implicitly specify that $X_E$ contains only variables that already appear in $X_H$.

of Army data, in which observations with zeros are simply deleted. Because the Army is the largest service, the truncation bias caused by deleting observations with zeros is smallest for the Army. We find that the elasticity estimates from the linear models are highly similar to those obtained in the truncated logarithmic analysis.)

The econometric results reported here do not include low-quality goals or performance. In the case of the Army, we found in preliminary analyses that low-quality goals and performance had the expected signs but were not statistically significant. In the case of the other services, quality goals were not set for much of the sample period.

Notice that (1a) intertwines recruiter effort and enlistee choices. If $X_H$ and $X_E$ contain common elements, as we contend they do, the coefficients on those variables are some $\gamma$ term plus some $\beta$ term; our data cannot disentangle the two. Similarly, changes in the coefficients on $M$, $P$, or $R$ may reflect changes in recruiting practices (i.e., effort), or they may reflect changes in how effort translates into successful recruiting. So although our econometric estimation can ascertain whether there has been a shift in the supply of high-quality recruits, it cannot ascertain whether the shift has been in supply as traditionally understood or in the effort expended on recruiting.

Despite the intertwining of recruiter effort and enlistee choice, the model does allow us to ask if an observed change in the model's coefficients is due solely to an altered response of potential recruits to recruiter effort. If only $\gamma_5$ has changed, then the $\beta$ terms will shift equiproportionately. If the $\beta$ terms change in some other fashion, then the observed change reflects either a change in the production of recruiter effort or changes in both effort production and the efficacy of that effort.

A simple analogy clarifies these distinctions. If we were studying the number of holes dug by ditch diggers, a sharp reduction in the number of holes dug could stem from any one of three causes: the ground might have frozen, the diggers might have exerted less energy, or the diggers might have been given spoons to replace their shovels. "The ground freezing" corresponds to potential recruits becoming less responsive to recruiter effort. "Diggers exerting less energy" corresponds to recruiters cutting back on their work effort.

"Replacing shovels with spoons" corresponds to reorganizations of recruiting activities in counterproductive ways (for example, sharply reducing the number of recruiting stations might prevent recruiters from being as successful as previously). A change in $\gamma_5$ would reflect the ground freezing. Other changes in the $\beta$ terms would reflect either diggers changing their effort expended or a change in their tools.

"Diggers changing their effort expended" could arise from either external circumstances or policy decisions that changed the incentives of recruiters. Apart from the Navy's recruiter incentive plan (see Asch (1990)), the incentive plans applied to the services' recruiters changed in relatively minor ways across the two time periods. However, the dramatic increase in the quality mix of recruits between the two periods might have affected recruiter attitudes toward an additional high-quality recruit, and hence their effort levels. Two policy changes that might have "replaced shovels with spoons" were the dramatic decrease in advertising between the two periods and the reduction in the number of recruiting stations in the later period. An external factor that might have "frozen the ground" was the drawdown itself, which might have given potential recruits and their families the impression that the military "wasn't hiring." Our econometric model cannot distinguish among these various hypotheses when the coefficients of the model do not change equiproportionately. So we must look to other evidence. (Other parts of the research project investigate why recruiter productivity might have changed. (See Orvis et al. (1996) and Oken and Asch (1997)).

What this research can tell us is (a) whether econometric models estimated with data from the 1980s predict the recruiting difficulties reported by the services in the 1990s; (b) whether the updated models predict recruiting difficulties in 1997; and (c) whether econometric models using more recent data give appreciably different estimates of the effects of economic and/or policy variables on high-quality recruiting, which might or might not confirm the predictions from (a).

## ECONOMETRIC STRATEGY

Our econometric strategy is designed to cope with six problems:

1. Biases may arise from omitted (and unobserved) policy variables.

2. Advertising expenditures, bonuses, and college fund opportunities are variables that are constant nationally and change only annually.

3. Ordinary least squares (OLS) estimators may be misleading when applied to data like ours, which have the structure of a cross-section of time-series.

4. Biases may arise from past contract performance appearing among the explanatory variables in the model.

5. Biases may arise from unemployment and pay being measured with error.

6. PUMAs vary widely in size, so we expect the disturbances in equation (1a) to be heteroskedastic.

Problems 1 and 2 are related. We noted above that if we include in our model dummy variables for every month or for every year in our sample, those dummies will be perfectly collinear with the advertising, bonus, and college fund variables. However, reducing the list of dummy variables—to include perhaps only dummies indicating the calendar month of the observation (to capture the fact that recruiting is higher in months close to school graduation)—risks biasing the coefficient estimates, owing to the effects of the unmeasured and therefore omitted policy variables. Basic to our econometric strategy is a two-stage approach: in the first, we include dummies for each month in the sample to unbiasedly estimate the effects of all but the advertising, bonus, and college fund variables; in the second, we include only one dummy for each calendar month to allow the estimation of coefficients for advertising, bonuses, and the college fund.

Problem 3 has two dimensions. First, from month to month the disturbances in (1a) display serial correlation. Second, PUMAs served by a single recruiting unit share some common experiences arising from the common command, giving rise to correlations between the disturbances in PUMAs that share a common recruiting unit. Both serial correlation and cross-PUMA correlations make OLS an inefficient procedure for estimating the coefficients of our model. More importantly, both these problems bias upward OLS estimates of the precision of our estimators (bias downward estimates of the standard errors of our coefficient estimates) and bias tests of hypotheses about the models' coefficients.

We use a feasible generalized least squares (FGLS) estimator to provide efficient estimates of the models' coefficients and to allow unbiased tests of hypotheses (Greene (1993), pp. 473–479). This procedure entails several steps, in which the degree of serial correlation and the degree of cross-PUMA correlation are both estimated and accounted for. (Cotterman (1986) was the first to use FGLS in the estimation of enlistment supply equations.)

Problem 4 complicates the FGLS procedure. Because of the serial correlation in the data, past performances, $P_H$ and $P_L$, are both correlated with the disturbance term in (1a); a surprisingly good (or bad) performance last month is indicative of another of the same this month. For FGLS to provide efficient coefficient estimates in such a case, the first step of the procedure, in which the degree of serial correlation is estimated, must use a special form of estimator called an "instrumental variables estimator." We do this.[2]

Problem 5 arises because our measures of unemployment and pay are based on relatively small samples and hence may suffer from measurement error. As with problem 4 above, instrumental variables estimation is a suitable strategy for coping with measurement error. We do this, using annually measured state-level unemployment rates and the numbers of qualified youths to construct a second measure of unemployment to use as an instrument.

Problem 6 requires that we weigh our observations to correct for the heteroskedastic in the disturbances. Since larger PUMAs will have more month-to-month variability in the number of contracts signed, we assume, for simplicity, that the variance of the disturbances is proportional to the PUMA's population. This in turn requires that we weigh the observations in inverse proportion to the square root of the PUMA's population size, to obtain efficient estimators. We also do this.

A seventh problem in estimating our model is that the demand-side variables, such as the number of recruiters assigned to a recruiting

---

[2]See Appendix B for details.

unit or the goals set for a recruiting unit, may depend upon the unit's past recruiting record. (We noted above that goals are set, in part, using models that rely on past performance.) Such a correlation between the demand-side variables and past performance would cause our coefficient estimates to be biased. The direction of these possible biases in recruiting resource variables' coefficients is unclear. Units that harvest a surprisingly high level of contracts are likely to spur some increase in their future goals, a correlation that is likely to bias upward the coefficient on goals. However, whether such a unit would be given more recruiting resources to grapple with their now higher goal or would lose resources to "more needy" units is unclear, making the direction of possible bias in the estimated recruiting resource variables' coefficients ambiguous. Berner and Daula (1993) do find some evidence that goals, in particular, are correlated with past performance in this way, but the extent of the bias caused was small. We do not account for these potential interactions, but we think the biases introduced are small.[3] (Had we uncovered more good candidates to use as instrumental variables, we would have countered these possible biases with the instrumental variables technique used to deal with problem 4.)

For a more detailed discussion of the stochastic structure of the disturbances in (1a) and of our estimation procedures, see Appendix B.

## PARTICULARS FOR THE FOUR SERVICES

The variables that appear in our econometric models differ from service to service. In the Army model, both high- and low-quality variables appear for both goal and past performance. For the Air Force and Navy, only a single measure of goals and of past performance is used for reasons discussed earlier. For the Marine Corps, we include no performance or goal variables because we were unable to obtain suitable data.

---

[3]If our FGLS estimator purges the disturbances of serial correlation, these biases disappear. Our estimates of the serial correlation coefficient change little across numerous specifications of the model, and our final estimates are little affected by relatively large changes in the estimated serial correlation coefficient used in the FGLS estimator. This finding, coupled with Berner and Daula's finding, leads us to think that recruiting resource endogeneity has little effect on our estimates.

However, some variables used are the same across services: the population variables, the pay variable, and the unemployment variable are the same for each service, and for each service we include the number of recruiters it has working in the PUMA. The population variables are three: the estimated number of high-quality high school graduates (in thousands) living in the PUMA (called High-Quality Available) and two similar variables for the numbers of black and Hispanic high-quality high school graduates. The civilian/military pay ratio variable is the average earnings of youths in the PUMA divided by the BPI; this is called Civilian/Military Pay. The unemployment variable is the number of unemployed persons (in thousands) between 16 and 64 in the PUMA. (Past studies, relying on logarithmic specifications, have used unemployment rates, rather than numbers of unemployed. In our linear specification, the number of unemployed and the population appear as separate variables.)

We estimate separate econometric models for the two periods FY83–FY87 and FY90–FY93. For this reason, we report in Table 1 the means and standard deviations of all the variables used in the models for each of these periods.

Table 1

Summary Values for Variables in High-Quality Recruit Supply Equations

| | All | | FY83–FY87 | | FY90–FY93 | |
|---|---|---|---|---|---|---|
| | Mean | Std. Dev. | Mean | Std. Dev. | Mean | Std. Dev. |
| **Common variables** | | | | | | |
| Civilian/Military Pay | 781.68 | 84.12 | 784.96 | 74.62 | 778.79 | 95.16 |
| Unemployed[a] | 8.94 | 16.03 | 9.51 | 16.29 | 8.41 | 15.77 |
| High-Quality Available (All)[a] | 3.09 | 4.41 | 3.24 | 4.63 | 2.96 | 4.20 |
| HQA (Black)[a] | 0.10 | 0.39 | 0.09 | 0.34 | 0.12 | 0.44 |
| HQA (Hispanic)[a] | 0.08 | 0.32 | 0.08 | 0.31 | 0.08 | 0.32 |
| **Army** | | | | | | |
| Gross HQ Contracts | 5.53 | 7.24 | 6.44 | 8.17 | 4.71 | 6.15 |
| Recruiters in PUMA | 5.24 | 7.40 | 5.29 | 7.48 | 5.21 | 7.33 |
| HQ Goal | 106.53 | 31.98 | 109.36 | 34.05 | 103.94 | 29.72 |
| LQ Goal | 67.13 | 35.30 | 82.60 | 29.02 | 53.00 | 34.60 |
| HQ Performance (lagged) | −19.93 | 38.01 | −8.01 | 35.86 | −30.81 | 36.62 |
| LQ Performance (lagged) | −8.20 | 30.34 | −1.23 | 30.14 | −14.55 | 29.10 |
| **Navy** | | | | | | |
| Gross HQ Contracts | 4.03 | 5.30 | 4.14 | 5.41 | 3.94 | 5.19 |
| Recruiters in PUMA | 3.81 | 5.38 | 3.47 | 4.93 | 4.13 | 5.74 |
| Goal | 107.51 | 93.13 | 99.09 | 100.90 | 115.19 | 84.69 |
| Performance (lagged) | 29.34 | 102.42 | 46.94 | 108.51 | 13.27 | 93.68 |
| **Air Force** | | | | | | |
| Gross HQ Contracts | 2.01 | 2.76 | 2.39 | 3.22 | 1.66 | 2.20 |
| Recruiters in PUMA | 1.37 | 1.97 | 1.56 | 2.26 | 1.21 | 1.64 |
| Goal | 74.86 | 62.05 | 89.86 | 73.59 | 61.16 | 45.04 |
| Performance (lagged) | 8.96 | 78.22 | 16.16 | 90.85 | 2.38 | 63.86 |
| **Marine Corps** | | | | | | |
| Gross HQ Contracts | 2.10 | 3.25 | 2.08 | 3.31 | 2.12 | 3.20 |
| Recruiters in PUMA | 2.62 | 3.94 | 2.62 | 4.17 | 2.61 | 3.71 |

[a]In thousands.

# Chapter Four

# EMPIRICAL RESULTS

We present our empirical results in three steps. First, we present the estimated coefficients and standard errors of estimates for the econometric models of high-quality enlistment supply based upon data from FY83–FY87 and FY90–FY93. Second, we provide two assessments of those estimates, one an examination of how the signs of the estimated coefficients accord with prior expectations, and the other a comparison with past econometric studies. Finally, we present the statistical and practical assessments of our econometric results: What do our models imply about the adequacy of high-quality enlistment supply in fiscal year 1997?

## ESTIMATED MODELS

We conducted our estimations in four stages. First, we estimated the model by using monthly dummy variables in the model, with no advertising, bonus, or college fund variables. In this stage we allowed the coefficient on each explanatory variable to differ between time periods. We used the results from these regressions to determine which variables had effects that changed between the two periods. In the second stage, we reestimated the model, again with monthly dummies, assuming that those variables with insignificant coefficient changes between periods indeed had no changes between periods. In the third stage, we replaced the monthly dummies with a dummy variable for each month of the calendar year (January, February, etc.,) and a dummy distinguishing the early period from the late. In this stage we carried forward and took as given the estimated coefficients from the second stage, and we estimated the effects of advertising, enlistment bonuses, and the Army College Fund, allowing the effects of each of these to differ between time periods. We used the results from these regressions to determine which variables had effects that changed between the two periods. In the fourth stage, we reestimated the models of stage three assuming that those variables with no significant coefficient changes between periods indeed had no changes between periods. Tables 2–5 present the final estimated models for the four services. Since much of our interest is in how the estimated models of high-quality recruit supply equations might have changed between the 1980s and the 1990s, we structured our estimation and report our results in a form that highlights such changes.

The chief econometric results for the four services are contained in Tables 2, 4, 6, and 8. These tables present the combined second-stage and fourth-stage estimates of the coefficients of the explanatory variables based on the sample period FY83–FY87, and estimates of the *changes* in those coefficients when estimated using data from FY90–FY93. (The estimates for advertising, enlistment bonuses, and the Army College Fund are obtained in the fourth stage, and all other coefficient estimates are drawn from the second stage.) The $t$-statistics reported for the earlier period are $t$-statistics for the coefficients themselves. The $t$-statistics reported for the latter period are the $t$-statistics for the estimated changes in the coefficients. Thus, in Table 2, which presents results for the Army, the first row, "High-Quality Available," reports the estimated coefficient (.22) for the total high-quality high school graduate variable for the period FY83–FY87; the next row labeled "High-Quality Available" reports how much the estimate of the coefficient changed when we used data from FY90–FY93 (.05). The $t$-statistic on the first row pertains to a test of the hypothesis that the coefficient on HQA is zero. The $t$-statistic on the subsequent row is pertinent to the null hypothesis that the coefficient on HQA was the same in the two periods. (Appendix B contains the coefficient estimates and $t$-statistics from estimation stages one and three.)

Efforts to analyze the effects of one service's recruiting resources, for example the number of recruiters, on another service's enlistments foundered on problems of multicollinearity.

## The Army Model

The estimated model for the Army, presented in Table 2, conforms well to prior theoretical expectations. Higher civilian pay or lower military pay deters high-quality enlistments. Greater unemployment spurs high-quality enlistments. Having more recruiters increases high-quality enlistments, while past success in meeting high-quality goals decreases such enlistments. Having more high-quality goals leads to greater Army high-quality enlistments. Enlistment bonuses and the Army College Fund increase high-quality Army enlistments. Theory does not tell us what differences there might be among potential recruits by race or ethnicity, but empirically we find that highly qualified blacks join the Army more often than similar whites, and whites more often than Hispanics; these differences became attenuated in the later period. We found no significant or appreciable effect of low-quality goals or past low-quality performance on high-quality enlistments.

We specified enlistment bonuses and the Army College Fund to have equal coefficients because we chose comparable measures for these two benefits: namely, the current discounted value of the offered benefits averaged across all specialties. Preliminary estimation that allowed separate coefficients for each variable revealed no statistically significant difference between the separately estimated coefficients in either time period; however, the standard errors were large in those estimations because the two variables have considerable positive covariation.

Apart from the already noted attenuation of racial differences, the chief differences between time periods for the two models are in the effects of recruiters, high-quality goals, and past high-quality recruiting performance. Additional recruiters and higher high-quality goals affect enlistments less during the later period, and successful past high-quality recruiting performance acted as a greater drag on high-quality enlistment.

A comparison of the estimates in Table 2 with past studies of enlistment behavior (see Warner and Asch (1995) for a more detailed discussion of the variation of estimates across past studies) raises questions that need addressing. Table 3 reports elasticity estimates drawn from other studies and compares them with our estimates.

## Table 2

### Estimated Supply Equations for High-Quality Army Recruits

| Variable | Estimate of Early-Period Coefficient | t-statistic |
|---|---|---|
| High-Quality Available | 0.22 | 8.29** |
| Black HQ Available | 4.00 | 12.00** |
| Hispanic HQ Available | -5.49 | -14.34** |
| Civilian/Military Pay | -0.12E-02 | -3.55** |
| Unemployed | 0.10 | 7.43** |
| Recruiters in PUMA | 0.77 | 31.88** |
| Past HQ Goal Success | -0.65E-02 | -14.61** |
| HQ Goal | 0.75E-02 | 7.28** |
| Enlistment Bonus & ACF | 0.38E-01 | 5.06** |

| Variable | Estimate of Change Between Periods (Late coefficient − early coefficient) | t-statistic |
|---|---|---|
| High-Quality Available | 0.05 | 1.34 |
| Black HQ Available | -2.89 | -6.29** |
| Hispanic HQ Available | 2.41 | 4.99** |
| Recruiters in PUMA | -0.24 | -9.61** |
| Past HQ Goal Success | -0.25E-02 | -4.09** |
| HQ Goal | -0.52E-02 | -3.72** |

Other coefficients not statistically different

| | | |
|---|---|---|
| Estimated serial correlation coefficient | 0.53** | .54** |
| Estimated variance of PUMA error component | 0.40E-02** | 0.41E-02** |
| Estimated variance of recruiting unit error component | 0.11E-03** | 0.16E-03** |

94,744 observations    F-statistic: 277.2** | 381.4**

*Significant at .05 level.
**Significant at .01 level.
NOTE: Model contains state and calendar month dummies, a dummy for the later period, and a dummy for months in which there were no Army College Fund benefits offered.

### Table 3

### Estimated Elasticities of High-Quality Recruit Supply with Respect to Army Supply Factors

| Factor | Past Studies | Present Study 83–87 | Present Study 90–93 |
|---|---|---|---|
| | | Estimate for the 1980s | |
| Youth Population | .24 | .08 | .16 |
| Recruiters | .60 | .51 | .60 |
| Military Pay | .55 | .19 | .31 |
| Unemployment Rate | .94 | .11 | .16 |
| Army College Fund | .17 | .07 | .01 |
| Enlistment Bonus Program | .07 | .08 | .003 |
| High-Quality Goal | .22 | .16 | .08 |

SOURCES: Asch and Orvis (1994, p. 31) and Polich, Dertouzos, and Press (1986).

The older estimates were those used in Asch and Orvis (1994) (except for the High-Quality Goal elasticity, which is taken from Polich, Dertouzos, and Press (1986)); the estimates calculated for this study are based on the coefficient estimates of Table 2, and are medians of the elasticities calculated for each observation. (Medians are used to avoid problems arising from zero enlistments sometimes appearing in denominators. Elasticities calculated at the ratio of sample means would differ little from these.) With the exceptions of highly qualified youths available and recruiters, our estimated elasticities are generally lower than those found elsewhere in the literature; our estimated youth and recruiter elasticities are close to those found by others.

The source of these differences is not in the linear specification of the model. We use the linear specification because numerous PUMAs with zero enlistments in a month make using a logarithmic model impossible. Appendix C shows that a logarithmic specification for the Army that deletes the observations for which there are zero enlistees yields elasticities highly similar to those obtained for the linear specification.

We experimented with including advertising, as measured by total advertising spending in a given year, in our models. The estimated coefficients were implausibly large (due, we think, to the inability of the aggregate measure to capture geographic differences in the impact of advertising spending). We therefore chose to delete advertising from the models. The coefficients of other variables were not much affected by the inclusion of advertising, but the prediction errors in specific periods were made larger as the aggregate trends in advertising moved with the aggregate trends in recruit levels.

Our estimates of the effects of enlistment bonuses and college funds are also based only on temporal variation, but here the variation is quarterly, offering us a better prospect of measuring the effects of these programs. Our estimates are not far from those of others, and we think them reliable.

We are hesitant about our estimates of unemployment elasticities because they are so much lower than what others have found. We had initially hoped that using substate unemployment data would provide us better estimates of unemployment's effects than had been previously obtained. We now worry that the benefits of more appropriate geographically based measures may have been outweighed by the costs of greater measurement error.

We are most confident in our estimates of the effects of recruiters and high-quality goals, elasticity estimates in close accord with what past researchers have found. These estimates are based on counts of recruiters and reports of performance and goals that should be fairly accurate, and for which we have considerable geographic variation. Moreover, when one allows the estimates for these variables to be influenced by temporal variation, the estimates do not change much at all.

Examination of Table 2 suggests a marked and significant change in the effects of variables related to recruiter effort. The marginal effect of additional recruiters on enlistments is one-third lower in the later period than in the earlier period. The marginal effect of an increase in the high-quality goal is two-thirds lower in the later period than in the earlier. And the marginal feedback effect of goal success on recruiter effort was more than one-third higher in the later period than in the earlier. These effects are not consonant with a simple reduction in the responsiveness of recruits to recruiter effort; such an explanation would require that the effects of all three effort-related variables shrink in the same proportion, which they do not. (An examination of the logarithmic model in Appendix C also shows changes between periods in the effort-related variables, although the elasticity for the recruiter variable per se does not change between periods.)

What are we to make of the changes in the recruiter-effort-related variables? There are two chief competing interpretations. The first is that structural changes have occurred in the use of recruiting resources and in the responsiveness of potential recruits to recruiter efforts. (The recruiters were given spoons to replace shovels and the ground has frozen, in the prosaic terms given above.) In this view, changes in Army advertising, management, or placement practices for recruiters have reduced the efficiency with which recruiting resource are applied, and changes in potential recruits' preferences and circumstances have made them less responsive to recruiters efforts. The second interpretation argues that recruiters conserved on effort ("the diggers expended less energy"). These two competing views call for markedly different uses of the early 1990s coefficients when forecasting the future. If the altered coefficients for the early 1990s are the product of recruiters easing up on effort, the stiffened recruiting requirements of the late 1990s should restore earlier behavior and the earlier period's coefficients. However, if the parameter changes are attributable to policy changes by the services and taste changes among the recruits, then no automatic readjustment of the coefficients will occur, and the early 1990s should serve as the basis for forecasting the late 1990s.

What, then, do we learn from our Army model about the adequacy of high-quality enlistment supply in FY97? Let us use FY93, the last year in our sample, to construct an analysis. In FY93, our data indicated that gross high-quality contracts numbered 46,071; our model predicted 52,198 gross high-quality enlistments for that year. The actual Army goal for gross high-quality contracts in FY93 was roughly 60,000. The Army's goal for FY97 was approximately 134,300 gross contracts, of which approximately 64 percent, or 86,000, would need to be high-quality contracts.

How likely is it that the Army would achieve this increased high-quality enlistment level of some 45 percent above 1993 levels? We make the projections without incorporating the resource changes and accession mission changes the Army made in 1997, in order to determine the extent of recruiting problems the Army would face without these changes. Thus, several variables in our models were inconsequentially different from their 1993 levels: the Army College Fund, enlistment bonuses, the military/civilian pay ratio, and high-quality youth populations. Unemployment, high-quality goals, numbers of recruiters, and service advertising would, however, all change substantially between 1993 and 1997. Unemployment was predicted to be down 22 percent, falling from 7 percent to 5.4 percent, which was close to the actual change of 28 percent. High-quality contract goals were projected to be up about 45 percent, which did not change during FY97 although the accession mission did. The number of recruiters would be down 11 percent or up as much as 30 percent, depending upon the reception of the Army's request at the time for an additional 2,200 recruiters in 1997. Advertising would be up 120 percent in real terms. (For purposes of generating forecasts, we add to these predictions about other variables the assumption that recruiter success in meeting high-quality goals in 1996 is essentially the same as in 1992.)

The Army model predicts that the 22 percent fall in unemployment would decrease high-quality recruits by 2,264. The 45 percent increase in goal, on the other hand, would lead to an increase in expected recruits of 1,278. A fall in recruiters of 11 percent would lead to a fall in recruits of 2,722, while a 30 percent increase in recruiters would lead to an increase of 7,733 recruits. The negative serial correlation that we estimated in the disturbances of the recruiting outcomes over time predicts that 8 percent, or 590, of the 6,127 difference between actual and predicted recruits in 1993 would persist into 1997. Summing these effects, one finds that the predicted levels of high-quality recruits would be 4,298 recruits lower than in 1993 if the number of recruiters falls by 11 percent, or 6,157 recruits higher than in 1993 if the number of recruiters is 30 percent higher than in 1993.

Since our Army model does not include advertising, some external prediction of the effect of markedly higher advertising expenditures is needed to complete one's forecast of high-quality recruits in 1997 vis-à-vis 1993. Drawing upon past literature, we choose an elasticity of .07 for advertising's effect on high-quality recruits (Asch and Orvis, 1994, p. 31). Applying this elasticity to the levels of predicted recruits obtained from our Army model under the two scenarios above (an 11 percent decline in recruiters and a 30 percent increase in recruiters), higher advertising would add 2,718 recruits over 1993 levels in the former case and 3,331 additional recruits in the latter case.[1]

Combining these effects of changes in the determinants of high-quality recruits yields forecasts of high-quality recruits for 1997. If the number of recruiters in 1997 was to be 11 percent below 1993 levels, we forecasted the number of high-quality recruits in 1997 to be 10 percent above the 1993 level; if the number of recruiters was raised to 30 percent above 1993 levels, our forecast of the number of high-quality recruits in 1997 is about 34 percent above the 1993 level. In the former case, the Army is forecasted to face a substantial recruit shortfall from the target of 86,000; in the latter case, the shortfall shrinks to a much smaller amount. The latter is well within the range of observed prediction errors of the model, whereas the former is definitely not. Hence, we conclude that the Army would likely face a substantial shortfall in high-quality recruits if additional recruiters or other resources are not increased. This conclusion assumes that at least for the immediate term, recruiter productivity will remain at its FY90–93 level, which we estimated to be about 25 percent lower than its FY83–89 level. Conceivably, improvements in recruiter management can increase recruiter productivity, which would further improve the Army's chance of making its FY97 goal.

The recruitment supply coefficients from the pre-1990s period applied to the variable values of the 1990s predicted markedly higher recruit levels (on average 10,000 higher per year) than did the 1990s coefficient estimates. This finding confirms the finding of Asch and Orvis (1994) that 1980s models would have predicted no recruit

---

[1]We get two estimates of the effect of advertising because a constant elasticity for advertising implies that an increase in advertising spurs a proportional increase in recruiting; hence, the effect of advertising depends upon the level of recruits implied by the other variables. We computed advertising's effects by computing

$$(\text{Army model prediction}) * e^{.07 * (\ln (2.2))},$$

where the 2.2 captures the higher level of service advertising in 1997 over 1993.

shortfalls in the 1990s. An implication of this finding is that if the altered supply responses estimated here from early-1990s data reflect only a temporary shift peculiar to the transition through the end of the Cold War and the implementation of the drawdown, the forecasted difficulties in meeting increased goals in 1997 would not appear. If supply conditions reverted in 1997 to earlier patterns as the forces emerged from the drawdown period, Army recruit supply would be adequate even with no increase in recruiters over 1993 levels. However, to the extent that the shift in supply is linked not to the drawdown per se but to the end of the Cold War, we do not expect supply conditions to revert to their earlier patterns. In any case, the importance of obtaining an adequate supply of high-quality recruits suggests that one should take seriously the likelihood that 1997 recruit supply patterns would be similar to those observed in the mid-1990s.

Consistent with our forecast, the Army experienced recruiting difficulties in the first half of FY97. In response, several steps were taken to increase recruiting resources to avoid a shortfall in high-quality recruits in 1997. Army College Fund benefits were increased from $30,000 to $40,000 for a four-year enlistment in critical skills, and enlistment bonuses were increased from $8,000 to $12,000 for high-quality enlistments in critical skills. The Army's advertising budget was also increased, and, importantly, the FY97 accession mission was reduced from 89,700 to 85,982.

## The Marine Corps Model

The estimated model for the Marine Corps, presented in Table 4, conforms well to prior theoretical expectations. Higher civilian pay or lower military pay deters high-quality enlistments in the earlier period, but not in the later period. (The estimate used here is for no effect of pay. The actual econometric result from stage one is a perverse effect. The actual estimate and its *t*-statistic appear in Appendix B.) Greater unemployment increases high-quality enlistments. More recruiters increase high-quality enlistments. Since we were unable to get reliable historical goal data for the Marine Corps, we do not include that in the model. Theory does not tell us what differences there might be among potential recruits by race or ethnicity, but empirically we find that highly qualified blacks join the

## Table 4

## Estimated Supply Equations for High-Quality Marine Corps Recruits

| Variable | Estimate of Early-Period Coefficient | t-statistic |
|---|---|---|
| High-Quality Available | 0.20 | 20.346** |
| Black HQ Available | 0.47 | 3.68** |
| Hispanic HQ Available | -0.43 | -3.12** |
| Civilian/Military Pay | -0.41E-03 | -3.51** |
| Unemployed | 0.05 | 12.48** |
| Recruiters in PUMA | 0.28 | 16.03** |
| | Estimate of Change Between Periods (Late coefficient – early coefficient) | t-statistic |
| High-Quality Available | -0.36E-02 | -0.24 |
| Black HQ Available | -0.08 | 0.49 |
| Hispanic HQ Available | -0.862 | -4.71** |
| Recruiters in PUMA | 0.09 | 4.85** |
| Civilian/Military Pay Unemployed | 0.41E-03*** | *** |
| | Not significantly different | |
| Estimated serial correlation coefficient | | 0.32**\|0.32** |
| Estimated variance of PUMA error component | | 0.11E-02**\| |
| | | 0.11E-02** |
| Estimated variance of recruiting unit error component | | -0.40E-05**\| |
| | | -0.63E-05** |
| 94,744 observations | F-statistic: 369.2**\|561.1** | |

*Significant at .05 level.
**Significant at .01 level.
***Variable constrained to zero; estimated coefficient would violate economic theory; see Appendix B for actual estimate and t-statistic.
NOTE: Model contains state and calendar month dummies, a dummy for the later period.

Marine Corps more often than similar whites, and whites more often than Hispanics; in the later period the difference between blacks and whites remained the same, but that between whites and Hispanics grew. Apart from the already noted perverse pay effect switch and the change in the white-Hispanic difference, the chief differences between time periods for the two models is in the effects of recruiters. Unlike the Army, additional recruiters affect Marine Corps enlistments more during the later period.

A comparison of the estimates in Table 4 with past studies of enlistment behavior again raises questions that need addressing. Like Table 3, Table 5 reports elasticity estimates drawn from other studies and compares them with our estimates, this time for the Marine Corps. The older estimates were those used in Asch and Orvis (1994); the estimates calculated for this study are based on the coefficient estimates of Table 4 and are medians of the elasticities calculated for each observation. With the exceptions of qualified youths and recruiters, our estimated elasticities are generally lower than those found elsewhere in the literature; our estimated youth and recruiter elasticities are close to those found by others.

As with the Army model, we are hesitant about our estimates of unemployment elasticities; even though the estimates for the Marine Corps are almost double those for the Army, they are much lower than what others have found.

### Table 5

**Estimated Elasticities of High-Quality Recruit Supply with Respect to Marine Corps Supply Factors**

| | Estimate for the 1980s | | |
| | Past | Present Study | |
| Factor | Studies | 83–87 | 90–93 |
|---|---|---|---|
| Youth Population | .24 | .35 | .34 |
| Recruiters | .60 | .53 | .62 |
| Military Pay | .55 | .33 | 0.0 |
| Unemployment Rate | .94 | .24 | .26 |

SOURCE: Asch and Orvis (1994, p. 31).

We are again most confident in our estimates of the effects of recruiters, estimates in close accord with what past researchers have found. In contrast to the Army, however, when one allows the estimates for these variables to be influenced by temporal variation, the estimated change in the recruiter effect shrinks and becomes statistically insignificant, though remaining positive in sign. So although we are very confident that the coefficient on recruiters did not fall for the Marine Corps as it did for the Army, we are less sure that the coefficient did in fact grow larger.

What do we learn from our Marine Corps model about the adequacy of high-quality enlistment supply in FY97? Let us use FY93, the last year in our sample, to construct an analysis. In FY93, gross high-quality Marine Corps contracts numbered 24,369; our model predicted 24,362 gross high-quality enlistments for that year. In 1997, the Marine Corps wanted approximately 35,312 net contracts, of which approximately 64 percent, or 22,600, would need to be high-quality contracts. In FY93, gross contracts were 21 percent higher than net contracts, so to achieve the desired 22,600 net contracts, the Marine Corps would need to recruit 27,346 gross high-quality contracts in 1997.

According to our model, what would be required to raise the predicted number of recruits from 24,362 to 27,346, an increase of approximately 12 percent? As a rule of thumb, our model suggests that for each 10 percent increase in recruiters, predicted high-quality Marine Corps enlistments will rise by 1,000. Thus, about a 30 percent increase in Marine Corps recruiters would be needed to raise predicted high-quality enlistments by 2,984. Even if the estimated effect of pay had not fallen, no reasonable pay increase would be likely to raise enlistments by 2,984. In short, our model suggests that the Marine Corps would face difficulty meeting its 1997 objective for high-quality recruits. On the other hand, our model does not include two important resource increases that occurred between FY93 and FY97 for the Marine Corps. Its college fund budget increased from about $1 million in FY93 to $11.5 million in FY97 (computed in FY97 dollars). The advertising budget also rose during this period by over 50 percent in real terms. These changes would help address the recruiting difficulties predicted by the model for FY97.

## The Air Force Model

The estimated model for the Air Force, presented in Table 6, conforms reasonably well to prior theoretical expectations. Higher civilian pay and lower military pay deter high-quality Air Force enlistments, but unemployment has no discernible effect on Air Force high-quality enlistments. More recruiters increase high-quality enlistments. Since Air Force goals, unlike the Army goals, are not explicitly tied to quality, theory does not tell us what the sign on the goal variable ought to be. We find that higher goals increased high-quality Air Force enlistments in the earlier period, and lowered them in the later period. Prior success in meeting goals leads to lower predicted high-quality enlistments. Theory also does not tell us what differences there might be among potential recruits by race or ethnicity, but empirically we find that potential black high-quality recruits join the Air Force more often than whites, and whites more often than Hispanics; in the later period these difference became attenuated.

Apart from the already noted changes in demographic effects, the chief differences between time periods are in the effects of recruiters, goals, performance, and pay. As with the Army, recruiters have a smaller effect in the later period. The effect of goals switches from positive in the early period to negative in the later period. Greater prior-goal success has a more negative effect on high-quality enlistments in the later period, as does higher civilian pay.

The estimation of the Air Force model differs from that of the other services. We found that accounting for correlations across PUMAs within recruiting areas in the Air Force data led to perverse, statistically significant signs for unemployment in both periods. Accounting for such correlation should lead to more efficient estimates of the model's coefficients, but ought not to induce such anomalous results. In the first stages of estimation, in which we use instrumental variables and correct for serial correlation, the coefficients on unemployment were small and statistically insignificant. We have chosen to rely on those estimates, forgoing the procedures we used with the other models, in which we estimated within-recruiting-unit correlations and used those correlations to reestimate the model.

## Table 6

### Estimated Supply Equations for High-Quality Air Force Recruits

| Variable | Estimate of Early-Period Coefficient | t-statistic |
|---|---|---|
| High-Quality Available | 0.16 | 14.42** |
| Black HQ Available | 1.45 | 10.36** |
| Hispanic HQ Available | -2.53 | -17.62** |
| Civilian/Military Pay | -0.43E-03 | -3.25** |
| Goal | 0.92E-03 | 3.58** |
| Past Goal Performance | -0.28E-03 | -2.32** |
| Recruiters in PUMA | 1.01 | 41.74** |
| | Estimate of Change Between Periods (Late coefficient – early coefficient) | t-statistic |
| High-Quality Available | -0.01 | -0.32 |
| Black HQ Available | -0.65 | 3.49** |
| Hispanic HQ Available | 0.97 | 4.81** |
| Civilian/Military Pay | -0.92E-02 | -4.00** |
| Goal | -0.030E-02 | -5.34** |
| Past Goal Performance | -0.17E-02 | -7.75** |
| Recruiters in PUMA | -0.26 | -6.27** |
| | | Step 2 | Step 4 |
| Estimated serial correlation coefficient | | 0.37** | 0.39** |
| 94,744 observations | F-statistic: 293.7** | 153.3** |

*Significant at .05 level.

**Significant at .01 level.

NOTES: The Air Force analysis was conducted without accounting for correlations across PUMAs within recruiting units. Model contains state and calendar month dummies, a dummy for the later period.

As with the Marine Corps, comparison of the Air Force estimates in Table 6 with past studies of enlistment behavior again shows our estimates for all but qualified youths and recruiters to be smaller than found in other studies. As in Tables 3 and 5, Table 7 reports elasticity estimates drawn from other studies and compares them with our estimates, this time for the Air Force.

We are again most confident in our estimates of the effects of recruiters, estimates in close accord with what past researchers have found. Moreover, as in the Army model, the decreased marginal effect of recruiters in the later period in Table 6 is robust across specification and estimation procedures. We are confident that the marginal effect of recruiters did fall during the early 1990s.

What do we learn from our Air Force model about the adequacy of high-quality enlistment supply in coming years? Let us once again use FY93, the last year in our sample, to construct an analysis. In FY93, gross high-quality Air Force contracts were 16,608; our model predicted 20,081 gross high-quality enlistments for that year, having underpredicted enlistments in each of the three previous years. In FY97, the Air Force wanted approximately 30,200 net contracts, 1,300 fewer than it obtained in 1993, and only 200 more than in the smallest recruiting class of the early 1990s, that of FY91. This modest need for recruits in FY97 suggests that the Air Force will have relatively little difficulty meeting its high-quality recruit objectives.

Table 7

**Estimated Elasticities of High-Quality Recruit Supply
with Respect to Air Force Supply Factors**

|  | Estimate for the 1980s | | |
|  | Past | Present Study | |
| Factor | Studies | 83–87 | 90–93 |
| --- | --- | --- | --- |
| Youth Population | .24 | .16 | .27 |
| Recruiters | .60 | .49 | .59 |
| Military Pay | .55 | .17 | 7.46 |
| Goal | .22 | .05 | –.15 |

SOURCES: Asch and Orvis (1994, p. 31), Polich, Dertouzos, and Press (1986).

## The Navy Model

The estimated model for the Navy, presented in Table 8, conforms least well to prior theoretical expectations. Both pay and unemployment have perverse effects. More recruiters do increase high-quality enlistments. Higher goals reduce predicted high-quality enlistments, a finding not inconsistent with theory, since Navy goals are not tied to quality. We find that highly qualified blacks join the Navy more often than whites, and whites more often than Hispanics. These findings raise some questions about the quality of our enlistment figures for the Navy, since the Navy itself, using its own enlistment figures, has not found similar anomalies.

Table 9, reporting elasticity estimates, is too sparse to merit much comment except that youth population and recruiter elasticities are once again close to what others have found.

Like the Air Force, the Navy does not face a steep increase in needed enlistments in FY97. Indeed, total accessions for FY97 are slated to be 6,500 *fewer* than in FY93. Adding recruiters is an effective tool for the Navy, relative to the fluctuations in its accession needs. A 5 percent increase in recruiters would raise predicted enlistments by 1,200. There were 20 percent more Navy recruiters in 1990 than in 1993.

## Table 8
### Estimated Supply Equations for High-Quality Navy Recruits

| Variable | Estimate of Early-Period Coefficient | t-statistic |
|---|---|---|
| High-Quality Available | 0.50 | 28.72** |
| Black HQ Available | 1.83 | 7.47** |
| Hispanic HQ Available | -2.16 | -8.55** |
| Civilian/Military Pay | -0.0002 | -0.77 |
| Goal | -0.13E-03 | -4.21** |
| Past Goal Performance | -0.35E-02 | -26.39** |
| Recruiters in PUMA | 0.59 | 49.13** |
| Unemployment | Wrong sign, set to zero | Significant*** |
| | Estimate of Change Between Periods (Late coefficient – early coefficient) | t-statistic |
| High-Quality Available | -0.08 | -3.33** |
| Black HQ Available | -0.41 | 1.25 |
| Hispanic HQ Available | 0.46 | 1.312 |
| Civilian/Military Pay | Wrong sign, set to zero | Significant*** |
| Unemployment | .01 | 3.80** |
| | Other coefficients not significantly different | |
| Estimated serial correlation coefficient | | 0.50**|.50** |
| Estimated variance of PUMA error component | | 0.24E-02**| |
| | | 0.25E-02** |
| Estimated variance of recruiting unit error component | | 0.34E-04**| |
| | | 0.56E-04** |
| 94,744 observations | | F-statistic: 227.964**|386.04** |

*Significant at .05 level.
**Significant at .01 level.
***See Appendix B for actual estimate and t-statistic.
NOTE:  Model contains state and calendar month dummies, a dummy for the later period.

## Table 9

### Estimated Elasticities of High-Quality Recruit Supply with Respect to Navy Supply Factors

| | Estimate for the 1980s | | |
| --- | --- | --- | --- |
| | Past | Present Study | |
| Factor | Studies | 83–87 | 90–93 |
| Youth Population | .24 | .33 | .24 |
| Recruiters | .60 | .42 | .53 |
| Goal | .22 | −.05 | −.06 |

SOURCES: Asch and Orvis (1994, p. 31), Polich, Dertouzos, and Press (1986).

# Chapter Five

# CONCLUSIONS AND RECOMMENDATIONS

In this report we estimate an econometric model of high-quality enlistment supply using data from two periods, FY83–FY87 and FY90–FY93. We found that the structure of high-quality enlistee supply has changed in the post–Cold War period, especially the effect of recruiters for the Army and the Air Force. What is not clear from the data is whether this shift is peculiar to the transitional period following the drawdown of forces and the end of the Cold War or will persist across the coming decade. We also confirm the result found in the preliminary analysis (Asch and Orvis, 1996) that supply in the mid-1990s should have been adequate to meet the services' demands for recruits. Finally, we found that an econometric model using 1990s data suggests that the Army would have difficulty meeting its recruiting goal for FY97. To address the recruiting problems it faced in FY97, the Army increased its recruiting resources, including educational benefits, enlistment bonuses, and advertising, and it lowered its accession mission. Consequently, the Army was able to meet its FY97 mission.

We estimated the model using geographically disaggregated data; the unit of observation is a PUMA, an area smaller than a state but generally larger than a county. Each PUMA is observed monthly over the periods FY83–FY87 and FY90–FY93. We used an instrumental-variables, feasible generalized least squares (FGLS) estimator that accounts for serial correlation in the disturbances and also for correlations in the disturbances within recruiting unit areas. The estimation procedure also accounted for correlation between lagged performance measures and the current disturbance in the gross contracts supply equation.

Gross non-prior-service contracts were the dependent variable in our models, a choice driven by data problems with the net contract variable. The close correlation between gross and net contracts in all the years for which we have good data on both indicates that the qualitative insights gleaned from using gross contracts apply also to net contracts.

We were surprised by two aspects of our results. First, geographic disaggregation to the PUMA level yielded less increase in efficiency than we had expected. Second, we had considerable difficulty estimating with confidence the coefficients on pay, bonuses, the Army College Fund and advertising, in contrast to the confidence we have in our results on goals and recruiters. The first surprise stems, we think, from the relatively poor quality of PUMA-level variables when compared with their more aggregated counterparts. The second surprise stems from our reliance on the variation in variables available in historical data. Bonuses and the Army College Fund have generally been changed in tandem, making it difficult to estimate their separate effects on enlistment behavior. Army pay has varied only annually and at the national level, providing relatively few observations for estimating its effects. Estimation of advertising's effects was hampered by the availability of only national expenditure levels.

The use of PUMA-level data greatly increased the time devoted to constructing and cleaning our data. It also markedly increased the computational burdens of the estimation process. The increased variation in geographic variables did not decrease standard errors by enough to warrant these added costs. Moreover, the PUMA-level variables were probably subject to more measurement error than recruiting-unit-level counterparts would have been.

Better estimates of enlisted supply will require better data. We recommend that the services consider experiments, akin to those conducted in the 1980s, to assess the effects of enlistment bonuses and educational benefits. Such experiments could insure independent variation in enlistment bonuses and education benefits, and hence allow a disentangling of their separate effects. Moreover, such experiments could also allow analysts to reduce the measurement errors that we think plague our local-level data. Analyses of individual recruiter's responses to alternative incentive schemes are also likely to need such experimental data.  Observations at the PUMA level are more appropriate to the study of individual recruiter behavior than are more aggregated observations, but our experience with such data suggests that experiments, with their potential for more variation in explanators and more attention to the quality of measurement, will be necessary if we are to make subtle inferences about individual recruiters' behavior.

Experiments with pay would be desirable as well but are less likely to be politically feasible.  Consequently, we urge analysts to consider how else pay and employment opportunities' effects might be estimated, with microdata sources being a possible alternative to the aggregated analysis attempted here.  We attribute our success in estimating the effects of recruiters and goals to the rich cross-sectional variation in those variables.  Another RAND research project is already undertaking the task of gathering geographically disaggregated advertising data to better estimate advertising's effects; we believe this approach has great merit.

# A FORMAL MODEL OF ENLISTMENT SUPPLY

## ESTIMATING HIGH-QUALITY ENLISTMENT SUPPLY

Ever since Working's classic 1927 article, "What Do Statistical Demand Curves Show?" econometricians have understood that the estimation of demand and supply curves requires disentangling one from the other. Early papers on the supply of high-quality recruits, however, argued that this task was particularly simple in the case of the supply of high-quality recruits. "The number of high-quality recruits is supply determined," wrote Charles Brown (1985, p. 228). By this he meant that the services accept as many high-quality recruits as are willing to join the military, so that observed variations in the number of high-quality recruits reflect movements in supply factors, not any change in demand conditions. Absent from Brown's widely shared view is a recognition of the roles of military recruiters and advertising in determining the numbers of high-quality recruits.

Few recruits simply walk in the door of recruiting stations, ask for forms, and just sign up. Most of them enter the military after much effort on the part of military recruiters who identify prospective recruits, provide those prospects with information, and woo them with tales of the benefits of military life. Richard Fernandez (1982, p. 46) succinctly states the first important point about recruiting: "It is natural to assume that more recruiters in an area means more high-quality recruits." Fernandez also provides an early, insightful comment on the econometric problems that recruiting introduces into the supply story; he writes that "recruiters may be moved among areas in response to past recruiting performance. This will also bias, probably upward, the estimate of recruiter effectiveness" (p. 47).

Dertouzos (1985, pp. 9–10) looked still deeper into the relationship between recruiting and the supply of high-quality enlistments.

> Most studies model the number of high-quality enlistments as a function of exogenous economic variables, X, and recruiting resources, R. However, ... an increase in the number of low-quality recruits will take time and resources away from activities that would increase high-quality enlistments. It is not possible to estimate [high-quality supply] without some explicit modeling of demand or recruiter objectives.

Even if the military wants as many high-quality recruits as might be willing to join, identifying those recruits and getting them signed requires effort by recruiters, whose incentives and preferences must be accounted for if we are to consistently estimate the supply of high-quality recruits.

## A FORMAL MODEL

The following model captures the complexities of enlistment supply noted above. The model is liberally adapted from that used in Polich, Dertouzos, and Press (1986).

The model assumes potential recruits and recruiters maximize their individual utilities. The potential recruits choose to enter the military, enter the civilian work force, or continue their schooling, depending on the costs and benefits of each option and on individuals' attitudes toward schooling and military and civilian life.

The supply of high-quality recruits is specified as

$$H = \gamma_1 D + \gamma_2 S + \gamma_3 X_H + \gamma_5 E_H + \varepsilon_H, \tag{1}$$

in which $H$ is the number of high-quality recruits, $D$ is a vector of time period dummy variables, $S$ is a vector of dummy variables indicating state, $X_H$ is a vector of traditional determinants of high-quality supply (including civilian and military pay, unemployment, a measure of population size, military advertising, and, where applicable, measures of bonuses and college fund opportunities offered to recruits), and $E_H$ is a measure of recruiting effort devoted to garnering high-quality recruits. Since recruiter effort, $E_H$, is unobserved, some adaptation of this specification is needed before the model is estimable. (The units of measure for effort can be chosen so that $\gamma_5$ is equal to one.) The term $\varepsilon_H$ is for stochastic disturbance.

The specification of the supply of low-quality recruits is qualitatively the same as that for high-quality recruits:

$$L = \gamma_6 D + \gamma_7 S + \gamma_8 X_L + \gamma_{10} E_L + \varepsilon_L, \tag{1b}$$

in which $X_L$ is a vector of determinants of low-quality supply (which may or may not differ from those for high-quality supply), $E_L$ is a measure of recruiting effort devoted to garnering low-quality recruits, and $\varepsilon_L$ is a stochastic disturbance term.

The utility functions of recruiters are assumed to depend on how successful they are in meeting their assigned goals, the incentives given them for meeting their various recruiting goals, and their level of recruiting effort ($E_H + E_L$). The tradeoff between success and work effort may in turn depend on how available recruits are in a given area in a given time period. Consequently, recruiters' effort level may also depend on exogenous determinants, such as $X_H$ or $X_L$. Recruiters maximize their utility by choosing the levels of effort, performance, and rewards that best accord with their tastes, given the recruiting environment they face. Hence, recruiter effort can be specified as

$$E_H = \tau_1 + \tau_2 X_E + \tau_3 M_H + \tau_4 M_L + \tau_5(P_H) + \tau_6(P_L) + \tau_7 R + \varepsilon_1 \tag{2}$$

$$E_L = \tau_8 + \tau_9 X_{EL} + \tau_{10} M_H + \tau_{11} M_L + \tau_{12}(P_H) + \tau_{13}(P_L) + \tau_{14} R + \varepsilon_2, \tag{2b}$$

in which $X_E$ contains numerous influences that affect recruiters' choices of effort level. The elements of $X_H$ are obvious candidates for inclusion in $X_E$, since they are indicative of how hard or easy it is to attract recruits; variables capturing incentive plans like the Navy's Freeman Plan might also be included in $X_E$, but we do not do so here. $M_H$ and $M_L$ are the recruiters' high- and low-quality recruiting goals; $P_H$ and $P_L$ are the recruiters' performance vis-à-vis high- and low-quality goals (the differences between contracts and goals); R is the total number of recruiters; and $\varepsilon_1$ and $\varepsilon_2$ are stochastic disturbance terms.

When recruiters got a high number of recruits relative to their high-quality goals last month ($P_H = (H - M_H) > 0$), we expect them to slack off some from high-quality recruiting effort and to focus more on low-quality recruiting, and contrariwise when $P_L = (L - M_L) > 0$. Hence, we expect $\tau_5$ to be negative and $\tau_6$ positive.

When high-quality goals are high, we expect more effort to be expended on garnering high-quality recruits; and when low-quality goals are high, we expect effort to be drained away from seeking high-quality recruits. Consequently, we expect $\tau_3$ to be negative and $\tau_4$ positive.[1]

## ECONOMETRIC PROBLEMS

The system of equations (1), (1b), (2), and (2b) poses three econometric problems. First, some "demand-side" variables, such as goals, number of recruiters, bonuses, and benefits, may be endogenous; second, effort is not an observed variable; and third, correlations between some independent variables and the model's disturbances make ordinary least squares procedures biased.

The appearance of goals, $M_H$, $M_L$, and R on the right-hand side of (2) and (2b) makes it clear that "demand factors" must be accounted for in estimating the supply of high-quality recruits. Less clear is whether these variables are endogenous, with current goals and advertising and recruiter allocations determined by past recruitment

---

[1]Many of the past analyses of recruitment supply have focused on the fact that high- and low-quality recruits are not equally easy to entice into the service. In general, we expect high-quality recruits to be more difficult to obtain, i.e., we expect $\gamma_5 < \gamma_{10}$. These past studies have consequently implicitly used (1) and (1b) to determine the combinations of $H$ and $L$ that are possible for a given level of total effort. This can be done by (a) choosing the units of effort so that $\gamma_5$ is equal to one, (b) solving (1) for $E_H$ and (1b) for $E_L$, (c) adding $E_H$ and $E_L$ to form effort, $E$, and (d) solving that relationship for $H$ as a function of $L$ and $E$. It is this relationship that others often refer to as the "enlistment supply function." Had we followed this route, we would have replaced (1a) with an equation similar in form but for the inclusion of an extra term $\beta L$. The interpretation of the coefficients, however, would be quite different, reflecting both high- and low-quality recruit supply considerations.

outcomes. Berner and Daula (1993) offer evidence that goals, at least, are indeed endogenous. However, their results also imply that the degree of endogeneity is small, so that overlooking this variable is not likely to much bias one's results. In this report, we make no effort to control for the endogeneity of the demand-side variables.

The unobservability of effort can be handled straightforwardly by replacing effort in (1) and (1b) with the right-hand sides of (2) and (2b). (See, for example, Polich, Dertouzos, and Press (1986).) This leads to new equations for $H$ and $L$:

$$H = \beta_1 + \gamma_1 D + \gamma_2 S + \gamma_3 X_H + \beta_2 X_E + \beta_3 M_H + \beta_4 M_L$$
$$+ \beta_5 (P_H) + \beta_6 (P_L) + \beta_7 R + u_H \tag{1a}$$

$$L = \beta_8 + \gamma_1 D + \gamma_2 S + \gamma_3 X_H + \beta_9 X_E + \beta_{10} M_H + \beta_{11} M_L$$
$$+ \beta_{12} (P_H) + \beta_{13} (P_L) + \beta_{14} R + u_L, \tag{2a}$$

where $\beta_i = \gamma_5 * \tau_i$ and $u_H = \varepsilon_H + \gamma_5 \varepsilon_1$. (1a) is the basic econometric model of high-quality recruit supply that we use in this report. We rely on a linear specification, rather than the logarithmic specification that has been commonly used in the past, because for many PUMAs, no contracts are signed in some months, making a logarithmic specification impossible, since the logarithm of zero is undefined.

Dertouzos' theoretical development of relationships like (1), (1b), (2), and (2b) assumes that the recruiters choose effort and performance so as to maximize utility. In practice, however, recruiters must choose their effort levels before they know for sure what their performance levels will be; the recruiters don't know how many recruits they will attract into the service. So recruiters' effort depends not on their actual performance levels, but on their expected performance levels. Hence, the variables $P_H$ and $P_L$ are better interpreted as expected performance levels and not actual levels. Past researchers have used actual performance as a proxy for expectations, which is not unreasonable, since as a month unfolds, recruiters get to see their actual performance emerge. However, we choose to proxy expectations with performance lagged one month, which is also reasonable, since as the month begins, the recruiter is likely to see last month's performance as giving information about what this month's performance will be.

The dependence of performance on contracts implies that the right-hand side of (1a) contains a lagged dependent variable. In consequence, because the disturbances in the recruiting equations are serially correlated, least squares estimates of (1a) would be biased. We use instrumental variables to avoid this bias (see Appendix B for a discussion of the instruments).

Notice that (1a) intertwines recruiter effort and enlistee choices inextricably. If $X_H$ and $X_E$ contain common elements, as we contend they do, the coefficients on those variables are some $\gamma$ term plus some $\beta$ term; our data cannot disentangle the two. Similarly, changes in the coefficients on $M$, $P$, or $R$ may reflect changes in recruiting practices (i.e., effort), or they may reflect changes in how effort translates into successful recruiting. So, while our econometric estimation can ascertain whether there has been a shift in the supply of high-quality recruits, it cannot ascertain whether the shift has been in supply as traditionally understood or in the effort expended on recruiting. We offer no solution to this fundamental identification problem. (Past research has sidestepped this identification issue by assuming that $\tau_1$, $\tau_2$, $\tau_3$, and $\tau_4$ are all zero.)

# THE STOCHASTIC SPECIFICATION AND ESTIMATION PROCEDURES

## THE STOCHASTIC SPECIFICATION

We ask two fundamental questions in this report: Did 1980s models of enlistment supply predict the recruiting difficulties the services reported in the 1990s? Did the supply of high-quality recruits shift appreciably between the period FY83–FY87 and the period FY89–FY93? We also ask how much of the observed decline in high-quality recruits can be attributed to changes either in the traditional determinants of enlistment supply or in their coefficients. The stochastic specification we choose for the model accommodates these purposes.

Our econometric strategy is designed to cope with five problems:

1. Biases may arise from omitted (and unobserved) policy variables.

2. The variables of advertising expenditures, bonuses, and college fund opportunities are constant nationally and change only annually.

3. Ordinary least squares (OLS) estimators may be misleading when applied to data like ours, which have the structure of a cross-section of time-series.

4. Biases may arise from past contract performance appearing among the explanatory variables in the model.

5. PUMAs vary widely in size, so we expect the disturbances in equation (1a) to be heteroskedastic.

Problems 1 and 2 are intertwined. If we include in our model dummy variables for every month or for every year in our sample, those dummies will be perfectly collinear with the advertising, bonus, and college fund variables. However, reducing the list of dummy variables to include perhaps only dummies indicating the calendar month of the observation (to capture the fact that recruiting is higher in months close to school graduation) risks biasing the coefficient estimates due to the effects of omitted policy variables. Basic to our econometric strategy is a two-stage approach; in the first stage we include dummies for each month in the sample to unbiasedly estimate the effects of all but the advertising, bonus, and college fund variables; in the second stage we include only one dummy for each calendar month to allow the estimation of coefficients for advertising, bonuses, and the college fund.

We also assume that the mean of the disturbance term in the model varies by state, so we include state-level dummy variables in the model. The state-level dummy variables in (1a) are intended to capture persistent, unmeasured characteristics that vary with geographic region. Preliminary testing indicated that using PUMA-level dummies would not add significantly to the explanatory power of the model, so we settled upon state-level dummies to capture geographic variation. We allow the state-level dummies' coefficients to vary between the two time periods under study.

The use of monthly dummies has the added benefit of correcting simply for any nonstationary in the underlying data, so long as any nonstationary series are cointegrated across PUMAs (for example, if the populations of all PUMAs share a common stochastic trend, they are each nonstationary but the deviations from their mean in any one time period is a stationary variable).

Problem 3 has two dimensions. First, from month to month the disturbances in (1a) display serial correlation. Second, PUMAs served by a single recruiting unit share some common experiences arising from the common command, a commonality that gives rise to correlations between the disturbances in PUMAs that share a common recruiting unit. Both serial correlation and cross-PUMA correlations make OLS an inefficient procedure for estimating the coefficients of our model. More important, both these problems bias upward OLS estimates of the precision of our estimators (bias downward estimates of the standard errors of our coefficient estimates) and bias tests of hypotheses about the models' coefficients.

Indeed, with 120 months of data on 911 PUMAs, one must guard assiduously against spurious estimates of statistical precision. Ordinary least squares estimates using such data are likely to yield severely downward-biased estimates of the standard errors of one's estimates.

We use a feasible generalized least squares (FGLS) estimator to provide efficient estimates of the models' coefficients and to allow unbiased tests of hypotheses (Greene (1993), pp. 473–479). This procedure entails several steps, in which the degree of serial correlation and the degree of cross-PUMA correlation are both estimated and both accounted for. (Cotterman (1986) was the first to use FGLS in the estimation of enlistment supply equations.)

Problem 4 complicates the FGLS procedure. Because of the serial correlation in the data, past performances, $P_H$ and $P_L$, are both correlated with the disturbance term in (1a); a surprisingly good (or bad) performance last month is indicative of another of the same this month. For FGLS to provide efficient coefficient estimates in such a case, the first step of estimation, in which the degree of serial correlation is estimated, must use a special form of estimator called an "instrumental variables estimator." We do this.

The key to instrumental variables estimation is to find variables that are correlated with the troublesome variables (in this case $P_H$ and $P_l$) but are uncorrelated with the disturbances in (1a). Since $P_H$ and $P_L$ are recruiting-unit-level variables, and all the $X$ variables in (1a) are PUMA-level variables, the values of recruiting-unit-wide populations of high- and low-quality recruits are plausible instruments to use in the present instance. The low-quality population within the PUMA is another plausible candidate, since it may be indicative of the marginal cost of low-quality recruits in a PUMA and hence correlate with a recruiter's division of labor between low- and high-quality recruits. Finally, since the PUMA's share of a recruiting unit's recruiters is our recruiter variable, another plausible instrument is the total number of recruiters in the PUMA. We use two-stage least squares to construct an instrument from these several candidates.

The relative naturalness of recruiting-unit variables as instruments is one further advantage of our using PUMA-level data for our analysis. We note that the central purpose of using instrumental variables in our procedure is to obtain consistent estimates of the serial correlation coefficient, not to obtain specific estimates of the effect of past recruiting performance per se.[1]

Once we consistently estimate the degree of serial correlation in our first estimation step, and account for that correlation subsequently, we need not use instrumental variables in the subsequent estimation steps—once the variables are transformed to correct for serial correlation, the transformed $P_H$ and $P_L$ variables are (in large samples) uncorrelated with the transformed disturbances. Had we used current performance as our proxy for expected performance, we'd have had to use instrumental variables in each step of the estimation procedure.

Problem 5 requires that we weigh our observations to correct for the heteroskedastic in the disturbances. Since larger PUMAs will have more month-to-month variability in the number of contracts signed, we assume, for simplicity, that the disturbances are proportional to the PUMA's population. This in turn requires that we weigh the observations in inverse proportion to the square root of the PUMA's population size, to obtain efficient estimators. We also do this.

This is an ad hoc specification based on the intuition that sample sums have a variance proportional to the number of observations. If the chance of enlisting were the same everywhere, the number of recruits would indeed have a variance proportional to the number of youths in the enlistment pool.

---

[1]In practice, our estimates of the serial correlation coefficient were quite robust to using alternative candidates for instrumentation and to alternative specifications of the model itself. Moreover, our final coefficient estimates were not very sensitive to moderately large changes in the estimated serial correlation coefficient used. (However, making no correction for serial correlation would have made a dramatic difference in the coefficients on past performance. The biases we seek to correct using FGLS are, apparently, substantial.)

## FORMALIZATION

The above description implies that our disturbances have an error-components structure. In any given time period, there are fresh shocks to the system. Following Cotterman (1986), we will assume that each component of the fresh element of the disturbance in a new time period has the same autocorrelation coefficient, and that all the serial correlations are first order. This allows us to correct for serial correlation in each PUMA's time-series in the usual fashion done with individual econometric series. We assume in this first analysis that all PUMAs have the same serial correlation coefficient.

Cotterman combined serial correlation correction with a correction for component errors. We adapt his approach to include a correction for heteroskedasticity. Cotterman included three components in his errors. The first was a common national error in a given time period; the second was a common state-level component for each time period; the third was a component common to the four services in a given state in a given time period. (We analyze the four services independently in this report, so we discuss this last error component no further here.)

The alternative to error components is fixed effects. For example, we include fixed effects for states and time periods, in contrast to Cotterman. How does one choose between error-component models and fixed-effects models? There are two vantage points from which to ask this question. The first considers the process by which the data are generated; the second considers the consequences of misspecification in each instance.

The first vantage point looks at the sampling framework implicitly at work: Will the observed effect persist across samples, or will it likely take on new values in the next sample observed? From this vantage, a fixed effect for states seems warranted. In a next sampling, the states observed would be the same as in the present sample, and the traits of the state that give rise to the effect are likely to be quite persistent. From this same vantage, one would be inclined to treat both temporal and recruiting-unit effects as random effects. There is no reason to think that the next sample of years will be like the last, so there is little reason on this ground for making the temporal effect a fixed effect. Similarly, the recruiters and their commanders in a next sample are unlikely to be the same ones as seen in the present sample, and hence the effects of recruiting units are likely to be random across samples.

It is the second vantage that supports making the temporal effects fixed in our model. If the time-period-specific shocks are correlated with other variables in the model, those other variables' coefficient estimates would be biased if the time-period effects were treated as random, not fixed. Avoiding such biases is our reason for making the temporal effects fixed in our analysis. In contrast, there is little reason to think that any other variables in our model are correlated with the effects of specific recruiters and commanders, so the second vantage offers no reason for treating the recruiting-unit effects as fixed. Since treating effects as random, not fixed, improves efficiency when it is correct, we think it better to make the recruiting-unit effects random in our specification.

In formal terms, we write

$$u_{imt} = r * u_{imt(t-1)} + v_{imt}, \tag{3}$$

in which $r$ is a first autoregressive coefficient, $u_{imt}$ is the disturbance in the $i$th PUMA served by the $m$th recruiting unit in time period $t$, and

$$v_{imt} = k_{imt} * (w'_{mt} + w_{imt}), \tag{4}$$

in which $w'_{mt}$ is a white noise innovation specific to the $m$th recruiting unit in period $t$, $w_{imt}$ is a white noise innovation specific to the $i$th PUMA in the $m$th recruiting unit at time $t$, and $k_{imt}$ is the square root of the population in the $i$th PUMA of the $m$th recruiting unit. The variances of $w'_{mt}$ and $w_{imt}$ are assumed to be proportional to the population of the PUMA. (The intuition underlying (4) is that both PUMA-specific and recruiting-unit-specific effects alter every individual's probability of enlisting. The cumulative effect of either of these individual effects in a given PUMA will have a variance approximately proportional to the population of the PUMA. The constant $k$ is needed to reflect the fact that the recruiting-unit effect on individuals may affect different numbers of individuals in each PUMA. We will use the estimated population of high-quality high school graduates in a PUMA to weigh the data to correct for this heteroskedasticity.)

## The Estimation Procedure

To estimate a model with errors of the form (3) and (4) by FGLS requires several steps.[2]   First, one consistently estimates[3] the model with no correction for the correlation structure of the disturbances, but correcting for heteroskedasticity.   Second, one estimates $r$ using the residuals from the first stage.   Third, one transforms each variable's $x_{imt}$ in the model to be $z_{imt} = [x_{imt} - \hat{r} * x_{im(t-1)}]/k_{imt}$, where $\hat{r}$ is the estimate of $r$.   Fourth, one consistently estimates the model using the variables in the $z$ form, again correcting for heteroskedasticity that is built into the $z_{imt}$ just created.   Fifth, one uses the residuals from the fourth step to estimate $\sigma_{im}^2$ and $\sigma_m^2$, the variances of $w'_{mt}$ and $w_{imt}$.   Sixth, one uses the estimates of these variances to transform the $z$ data.   The transformation is specific to each recruiting unit.   Each $z_{imt}$ has subtracted from it a fraction of the mean of $z$ for the given time period, for the recruiting unit serving the PUMA.   The fraction, $f$, is

$$f = 1 - \left[\sigma_{im}^2 \big/ (\sigma_{im}^2 + T * \sigma_m^2)\right]^{0.5}. \tag{7}$$

Sixth, one consistently reestimates the model using the transformed $z$ variables, again correcting for heteroskedasticity.   Finally, one reestimates $r$ using the coefficient estimates from step six, and passes through all the other steps one more time.   This iterative procedure provides asymptotically efficient estimators of the model's coefficients, and these estimates can be used to conduct asymptotically valid $F$-tests of hypotheses about the model's coefficients.   (Since our corrections for heteroskedasticity are only approximate, the effi

---

[2]See Greene (1993, Chapter 16) for an extended discussion of feasible generalized least squares and of estimating error components models.

[3]Throughout, "consistent estimation" would mean OLS or weighted least squares if there were no endogeneity problems in the model.  Given the endogeneity of several variables in the model, instrumental variables is needed in lieu of OLS or weighted least squares.

ciency and the asymptotic validity of the tests are also only approximate.)

## ESTIMATION RESULTS FOR STAGES ONE AND THREE FOR EACH SERVICE

This section reports for each service estimation results from stages one and three of our multistage estimation procedure.

## Table B.1

## Estimated Supply Equations for High-Quality Army Recruits:  Stage 1

| Variable | Estimate of Early-Period Coefficient | t-statistic |
|---|---|---|
| High-Quality Available | 0.22 | 8.31** |
| Black HQ Available | 3.97 | 11.90** |
| Hispanic HQ Available | -5.69 | -13.98** |
| Civilian/Military Pay | -0.12E-02 | -3.56** |
| Unemployed | 0.12 | 6.61** |
| Recruiters in PUMA | 0.75 | 26.31** |
| Past HQ Goal Success | -0.71E-02 | -11.34** |
| HQ Goal | 0.87E-02 | 7.26** |
| | Estimate of Change Between Periods (Late coefficient – early coefficient) | t-statistic |
| High-Quality Available | 0.060 | 1.48 |
| Black HQ Available | -2.59 | -5.26** |
| Hispanic HQ Available | 2.97 | 5.05** |
| Recruiters in PUMA | -0.19 | -5.00** |
| Past HQ Goal Success | -0.20E-02 | -2.44** |
| HQ Goal | -0.65E-02 | -4.19** |
| Estimated serial correlation coefficient | | 0.19** |
| Estimated variance of PUMA error component | | 0.40E-02** |
| Estimated variance of recruiting unit error component | | 0.11E-03** |
| 94,744 observations | | F-statistic:  269.6** |

*Significant at .05 level.
**Significant at .01 level.
NOTE:  Model contains state and monthly dummies, and a dummy for months in which there were no ACF benefits offered.

Table B.2

**Estimated Supply Equations for High-Quality Army Recruits: Stage 3**

| Variable | Estimate of Early-Period Coefficient | t-statistic |
|---|---|---|
| High-Quality Available | 0.22 | . |
| Black HQ Available | 4.00 | . |
| Hispanic HQ Available | –5.49 | . |
| Civilian/Military Pay | –0.12E-02 | . |
| Unemployed | 0.98E-01 | . |
| Recruiters in PUMA | 0.77 | . |
| Past HQ Goal Success | –0.65E-02 | . |
| HQ Goal | 0.75E-02 | . |
| Enlistment Bonus & ACF | 0.28E-01 | 2.37** |
| Army Advertising | 0.26E-01 | 9.26** |
| Other Service Advertising | –0.80-02 | –3.41** |

| | Estimate of Change Between Periods (Late coefficient – early coefficient) | t-statistic |
|---|---|---|
| High-Quality Available | 0.05 | . |
| Black HQ Available | –2.89 | . |
| Hispanic HQ Available | 2.41 | . |
| Recruiters in PUMA | –0.24 | . |
| Past HQ Goal Success | –0.25E-02 | |
| HQ Goal | –0.52E-02 | |
| Army Advertising | –0.76E-02 | –2.11* |
| Other Service Advertising | –0.85E-02 | 0.24 |

| | | |
|---|---|---|
| Estimated serial correlation coefficient | | 0.25** |
| Estimated variance of PUMA error component | | 0.40E-02** |
| Estimated variance of recruiting unit error component | | 0.11E-03** |
| 94,744 observations | | F-statistic: 269.6** |

*Significant at .05 level.
**Significant at .01 level.

NOTE: Model contains state and calendar month dummies, a dummy for the later period, and a dummy for months in which there were no ACF benefits offered. "." indicates coefficient was imposed on this stage, having been estimated in a previous stage.

Table B.3

Estimated Supply Equations for High-Quality Marine Corps Recruits: Stage 1

| Variable | Estimate of Early-Period Coefficient | t-statistic |
|---|---|---|
| High-Quality Available | 0.20 | 20.32** |
| Black HQ Available | 0.46 | 3.64** |
| Hispanic HQ Available | −0.46 | −3.22** |
| Civilian/Military Pay | −0.42E-03 | −3.50** |
| Unemployed | 0.58E-01 | 10.52** |
| Recruiters in PUMA | 0.28 | 13.96** |
| | Estimate of Change Between Periods (Late coefficient − early coefficient) | t-statistic |
| High-Quality Available | −0.67E-02 | −0.44 |
| Black HQ Available | 0.95E-01 | 0.51 |
| Hispanic HQ Available | −0.79 | −3.79** |
| Recruiters in PUMA | 0.11 | 3.75** |
| Civilian/Military Pay | .004 | 2.72** |
| Estimated serial correlation coefficient | | 0.1E-01** |
| Estimated variance of PUMA error component | | 0.11E-02** |
| Estimated variance of recruiting unit error component | | 0.99E-06 |
| 94,744 observations | | F-statistic: 364.0** |

*Significant at .05 level.

**Significant at .01 level.

NOTE: Model contains state and monthly dummies, and dummy for the later period.

Table B.4

Estimated Supply Equations for High-Quality Air Force Recruits: Stage 1

| Variable | Estimate of Early-Period Coefficient | t-statistic |
|---|---|---|
| High-Quality Available | 0.16 | 13.99** |
| Black HQ Available | 1.46 | 10.38** |
| Hispanic HQ Available | -2.45 | -14.64** |
| Civilian/Military Pay | -0.44E-03 | -3.28** |
| Unemployed | -0.66E-02 | -0.93** |
| Recruiters in PUMA | 1.05 | 24.92** |
| Past Goal Success | -0.28E-03 | -2.32** |
| Goal | 0.96E-03 | 3.67** |

| | Estimate of Change Between Periods (Late coefficient – early coefficient) | t-statistic |
|---|---|---|
| High-Quality Available | -0.30E-02 | -0.18 |
| Black HQ Available | -0.55 | -2.58** |
| Hispanic HQ Available | 1.01 | 4.21** |
| Recruiters in PUMA | -0.25 | -3.70** |
| Past Goal Success | -0.17E-02 | -7.70** |
| Goal | -0.31E-02 | -5.43** |

| | | |
|---|---|---|
| Estimated serial correlation coefficient | | 0.34** |
| 94,744 observations | F-statistic: 290.1** | |

*Significant at .05 level.
**Significant at .01 level.
NOTES: The Air Force analysis was conducted without correcting for correlations across PUMAs within recruiting units. Model contains state and calendar month dummies, and a dummy for the later period.

## Table B.5

## Estimated Supply Equations for High-Quality Navy Recruits: Stage 1

| Variable | Estimate of Early-Period Coefficient | t-statistic |
|---|---|---|
| High-Quality Available | 0.50 | 25.53** |
| Black HQ Available | 1.90 | 7.59** |
| Hispanic HQ Available | −1.54 | −5.13** |
| Civilian/Military Pay | −0.24E-03 | −0.77 |
| Unemployed | −0.05 | −4.05** |
| Recruiters in PUMA | 0.68 | 23.81** |
| Past Goal Success | −0.34E-02 | −19.38** |
| Goal | −0.15E-02 | 3.80** |
| | Estimate of Change Between Periods (Late coefficient − early coefficient) | t-statistic |
| High-Quality Available | −0.07 | −2.35** |
| Black HQ Available | −0.26 | −0.70 |
| Hispanic HQ Available | 0.85 | 1.94** |
| Recruiters in PUMA | −0.66E-02 | −0.17 |
| Past Goal Success | −0.28E-03 | −1.02 |
| Goal | −0.73E-03 | 1.03 |
| Civilian/Military Pay | .01 | 3.80 |
| Estimated serial correlation coefficient | | 0.50** |
| Estimated variance of PUMA error component | | 0.24E-02**| |
| Estimated variance of recruiting unit error component | | 0.52E-04**| |
| 94,744 observations | | F-statistic:  218.20**| |

*Significant at .05 level.
**Significant at .01 level.
NOTE:  Model contains state and monthly dummies, and a dummy for the later period.

# ESTIMATION RESULTS FOR AGGREGATE AND
# LOGARITHMIC MODELS FOR THE ARMY

Past studies of recruiting supply have relied upon data aggregated to the state level or to the level of the military recruiting unit. Since some of our empirical results differ from those found in such aggregate studies, we ask here whether our results would have been different had we used more highly aggregated data. Looking in Table C.1 at the results of an aggregate analysis also allows us to assess the informational gains that we achieved by using more disaggregated data. (These analyses included advertising. Dropping advertising from these equations would yield very similar results.)

The coefficients in Table C.1 are not directly comparable in magnitude because of the differing levels of aggregation involved in the aggregate and disaggregate analyses. Nonetheless, the $t$-statistics are comparable. Larger $t$-statistics indicate more precise estimates of a variable's effects so long as the estimated overall effects of the variables are comparable when aggregation is accounted for. The results of Table C.2 support the claim that the estimated effects of variables are of similar magnitude between the aggregate and disaggregate analyses. Examination of the $t$-statistics does support the claim that the disaggregate analysis provides informational gains. The $t$-statistics in the disaggregated analysis are generally larger than in the aggregate analysis. Nonetheless, the gains are smaller than we anticipated. It is not clear that the gains were worth the increased costs of data preparation and estimation.

This report relies on a linear specification of the recruit supply model because in many PUMAs, in many months, there are no recruits.

Table C.1

Comparison of Aggregated and Disaggregated Results

| | Estimated Early-Period Coefficients | | | |
| | Aggregated Data | | Disaggregated Data | |
| Variable | Parameter Estimate | t-stat | Parameter Estimate | t-stat |
|---|---|---|---|---|
| High-Quality Available | 1.39 | 9.53 | 0.22 | 8.29 |
| Black HQ Available | 5.65 | 6.34 | 4.00 | 12.00 |
| Hispanic HQ Available | –13.44 | –14.42 | –5.49 | –14.34 |
| Recruiters in PUMA | 0.46 | 8.77 | 0.77 | 31.88 |
| Past HQ Goal Success | –0.07 | –5.74 | –0.01 | –14.61 |
| HQ Goal | 0.18 | 11.08 | 0.01 | 7.28 |
| Enlistment Bonus | 0.10 | 0.43 | 0.02 | 2.85 |
| Other Service Advertising | 0.29 | 6.09 | –0.01 | –3.44 |
| ACF | 0.10 | 0.43 | 0.02 | 2.85 |
| Army Advertising | 0.41 | 7.19 | 0.03 | 9.32 |
| Unemployed | 0.04 | 1.02 | 0.10 | 7.43 |
| Civilian/Military Pay | –0.02 | 2.20 | –0.001 | –3.55 |

**Table C.1—continued**

| Variable | Estimated Changes in Coefficients Between Early and Late Periods | | | |
|---|---|---|---|---|
| | Aggregated Data | | Disaggregated Data | |
| | Parameter Estimate | t-stat | Parameter Estimate | t-stat |
| High-Quality Available | -0.74 | -4.62 | 0.05 | 1.34 |
| Black HQ Available | -8.21 | -5.38 | -2.89 | -6.29 |
| Hispanic HQ Available | 8.45 | 6.19 | 2.41 | 4.99 |
| Recruiters in PUMA | -0.15 | -2.43 | -0.24 | 9.61 |
| Past HQ Goal Success | -0.09 | -4.39 | -0.002 | -4.09 |
| HQ Goal | -0.13 | -5.33 | -0.005 | -3.72 |
| Enlistment Bonus | 1.67 | 2.10 | Dropped as insignificant | |
| Other Service Advertising | -0.78 | -8.62 | -0.016 | -3.95 |
| ACF | 1.67 | 2.10 | Dropped as insignificant | |
| Army Advertising | 0.13 | 1.64 | -0.007 | |
| Unemployed | 0.13 | 2.52 | Dropped as insignificant | |
| Civilian/Military Pay | Dropped as insignificant | | | |

Such observations could not be used in algorithmic specification of the model. However, past, more highly aggregated studies have relied on logarithmic specifications of the recruit supply model. In this appendix, we compare the linear specification results with the results of a logarithmic specification from which observations with zero recruits are simply excluded.

Table C.2 compares the estimated elasticities (estimated at the median of sample values in the case of the linear models) for linear and logarithmic specifications for both the PUMA-level disaggregated data and the recruiting-unit-level aggregated data. The estimated elasticities are broadly similar across the four analyses.

Table C.2

Comparisons Among Aggregated/Disaggregated Results and Between Linear/Logarithmic Results

| Variable | Past Studies | Disaggregated Data | | | | Aggregated Data | | | |
|---|---|---|---|---|---|---|---|---|---|
| | | Log Model Elasticity | | Linear Model Impl. Elasticity | | Log Model Elasticity | | Linear Model Impl. Elasticity | |
| | | Early | Late | Early | Late | Early | Late | Early | Late |
| Youth Pop. | .24 | .11 | .11 | .08 | .16 | .56 | .54 | .77 | .42 |
| Recruiters | .60 | .46 | .48 | .51 | .60 | .54 | .43 | .40 | .36 |
| HQ Goal | .22 | .23 | .11 | .16 | .08 | .15 | .05 | .18 | .05 |
| Bonuses[a] | .07 | .04 | .04 | .05 | .01 | .09 | .03 | .01 | .01 |
| ACF[b] | .17 | .04 | .04 | .06 | .002 | .09 | .03 | .01 | .03 |
| Nat. Advert. | .07 | .79 | .23 | 1.05 | .94 | 1.06 | .23 | .83 | .58 |
| Unemployment | .94 | .32 | .23 | .11 | .16 | .12 | .05 | .06 | .27 |
| Military Pay[b] | .55 | .17 | .23 | .19 | .31 | .45 | .08 | .16 | .19 |

[a]In the linear model, bonuses and ACF had a single coefficient but differing elasticities. In the log model, this is flipped.
[b]The later-period income elasticity was imposed from without, as the estimated coefficient using both time and cross-series variation was perverse, as had been the case in the linear specification as well. No other coefficient estimate is much affected by this restriction.

Asch, Beth J., *Navy Recruiter Productivity and the Freeman Plan,* Santa Monica, CA: RAND, R-3713-FMP, 1990.

Asch, Beth J., and Lynn A. Karoly, *The Role of the Job Counselor in the Military Enlistment Process,* Santa Monica, CA: RAND, MR-315-P&R, 1993.

Asch, Beth J., and Bruce R. Orvis, *Recent Recruiting Trends and Their Implications: Preliminary Analysis and Recommendations,* Santa Monica, CA: RAND, MR-549-A/OSD, 1994.

Ash, Colin, Bernard Udis, and Robert F. McNown, "Enlistments in the All-Volunteer Force: A Military Personnel Supply Model and Its Forecasts," *American Economic Review,* Vol. 73, No. 1, March 1983, pp. 145–155.

Berner, J. Kevin, and Thomas Daula, "Recruiting Goals, Regime Shifts, and the Supply of Labor to the Army," *Defense Economics,* Vol. 4, No. 4, 1993, pp. 315–328.

Brown, Charles, "Military Enlistments: What Can We Learn from Geographic Variation?" *American Economic Review,* Vol. 75, No. 1, March 1985, pp. 228–234.

Cotterman, Robert F., *Forecasting Enlistment Supply: A Time-Series of Cross-Sections Model,* Santa Monica, CA: RAND, R-3252-FMP, July 1986.

Dale, C., and C. Gilroy, "Estimates in the Volunteer Force," *American Economic Review,* Vol. 75, 1985, pp. 547–441.

Daula, T., and D. Smith, "Estimating Enlistment Supply Models for the U.S. Army," in R. Ehrenberg, (ed.), *Research in Labor Economics*, Greenwich, CT: JAI Press, 1985, pp. 261–309.

Dertouzos, James N., *Recruiter Incentives and Enlistment Supply*, Santa Monica, CA: RAND, R-3065-MIL, May 1985.

Dertouzos, James N., and J. Michael Polich, *Recruiting Effects of Army Advertising*, Santa Monica, CA: RAND, R-3577-FMP, 1989.

Fernandez, Richard L., *Enlistment Effects and Policy Implications of the Educational Assistance Test Program*, Santa Monica, CA: RAND, R-2935-MRAL, September 1982.

Goldberg, Lawrence, *Enlisted Supply: Past, Present, and Future*, Alexandria, VA: Center for Naval Analyses, CNS 1168, 1982.

Greene, William H., *Econometric Analysis*, 2d ed., New York: Macmillan, 1993.

Hosek, James R., and Christine E. Peterson, *Enlistment Decisions of Young Men*, Santa Monica, CA: RAND, R-3238-MIL, July 1985.

Hosek, James R., and Christine E. Peterson, *Serving Her Country: An Analysis of Women's Enlistment*, Santa Monica, CA: RAND, R-3853-FMP, 1990.

Hosek, James R., Christine E. Peterson, and Joanna Zorn Heilbrun, *Military Pay Gaps and Caps*, Santa Monica, CA: RAND, MR-368-P&R, 1994.

Kilburn, M. Rebecca, Lawrence M. Hanser, and Jacob A. Klerman, *Estimating AFQT Scores for National Educational Longitudinal Study (NELS) Respondents*, Santa Monica, CA: RAND, MR-818-OSD/A, 1998.

Nelson, G., "The Supply and Quality of First-Term Enlistees Under the All-Volunteer Force," in W. Bowman, R. Little and G. T. Sicilia, (eds.), *The All-Volunteer Force After a Decade*, Washington, D.C.: Pergamon-Brasseys, 1986, pp. 23–51.

Oken, Carole, and Beth J. Asch, *Encouraging Recruiter Achievement: A Recent History of Recruiter Incentive Programs*, Santa Monica, CA: RAND, MR-845-OSD/A, 1997.

Orvis, Bruce R., Narayan Sastry, and Laurie L. McDonald, *Military Recruiting Outlook: Recent Trends in Enlistment Propensity and Conversion of Potential Enlisted Supply,* Santa Monica, CA: RAND, MR-677-A/OSD, 1996.

Polich, J. Michael, James N. Dertouzos, and S. James Press, *The Enlistment Bonus Experiment,* Santa Monica, CA: RAND, R-3353-FMP, April 1986.

Thomas, George W., and Linda Gorman, *Estimation of High-Quality Available and Interested,* Monterey, CA: Naval Postgraduate School, Department of Administrative Sciences, August 1991.

Warner, John T., and Beth J. Asch, "The Economics of Military Manpower," in Keith Hartley and Todd Sandler (eds.), *Handbook of Defense Economics,* vol. 1, New York: Elsevier Press, 1995, pp. 349–398.

Working, E., "What Do Statistical Demand Curves Show?" *Quarterly Journal of Economics,* Vol. 41, 1926, pp. 212–235.